The Automobile: A Century of Progress

Prepared under the auspices of the
SAE Historical Committee

SAE HISTORICAL COMMITTEE

INTERNATIONAL ®

Society of Automotive Engineers, Inc.
Warrendale, Pa.

Library of Congress Cataloging-in-Publication Data

The automobile: a century of progress/prepared under the
    auspices of the SAE Historical Committee.
            p.   cm.
    Includes index.
    ISBN 0-7680-0015-7
    1. Automobiles—Design and construction—History.
I. SAE Historical Committee.
TL240.A795   1997
629.2'31'09—dc21

Copyright © 1997 Society of Automotive Engineers, Inc.
            400 Commonwealth Dr.
            Warrendale, PA 15096-0001
            U.S.A.
            Phone: (412) 776-4841
            Fax: (412) 776-5760
            http://www.sae.org

ISBN 0-7680-0015-7

SAE Order No. R-203

Table of Contents

Foreword

Welcome to *The Automobile: A Century of Progress*! This book marks the culmination of a five-year effort by the SAE Historical Committee commemorating the 100th anniversary of the automobile industry in the United States.

When this project was first contemplated in 1992, we anticipated doing a chronological review of the technical history of the automobile in the "who-did-what-when" manner that is typical of most automotive history publications. Then, one of the committee members suggested, "Let's deviate from the norm and use the systems approach because this is the way a car is engineered." His suggestion was unanimously accepted. With inspired enthusiasm, we set out to develop a series of twelve systems-oriented articles for publication in *Automotive Engineering* magazine, delineating the technological progress of the American automobile over the past 100 years. After the task was defined, selecting the exact topics to be covered turned out to be a greater challenge than we had anticipated. In an effort to obtain the assistance of the various SAE technical committees, we aligned the subjects of our articles accordingly and then proceeded to enlist appropriate authors from the committees. However, as the clock ticked and the calendar rolled, we experienced mixed response from the committees. Thus, we came to what should have been an obvious realization: Technical committees look to the future and not to the past, which is as it should be.

After this bit of wisdom was understood, we refocused our attention on the more traditional automotive systems and sought authors from SAE at large to write about the topics on which they had both expertise and genuine historical interest. This turned out to be the proper approach. As a result, we produced and published a series of fourteen articles covering ten topics, including a time line in the September 1996 issue of *Automotive Engineering*. Those articles have now been compiled in this book. Although we are indeed proud of our efforts, our pride is tempered somewhat because we were unable to complete the series as originally intended. When we could not find an author available to do justice to driveline, chassis, and lighting systems development, we elected to eliminate these topics rather than cover them inadequately. Rest assured that we regret these omissions as much as you do.

Having said that, let us now extend *kudos* to all who have contributed to this book. We trust that you will enjoy the fruits of our effort as we pledge to continue serving SAE in the preservation of the history of mobility technology.

James K. Wagner, Coordinator
Centennial Publication Effort
SAE Historical Committee
April 2, 1997

Powerplant Perspectives:
Part I

Gordon L. Rinschler and Tom Asmus
Chrysler Corporation

In 1824, Sadi Carnot concluded his solitary treatise on heat engines, which became a significant element in the foundation of thermodynamics, with the following:

> "We should not expect ever to utilize in practice all the motive power of combustibles. The attempts made to attain this result would be far more hurtful than useful if they caused other important considerations to be neglected. The economy of the combustible is only one of the conditions to be fulfilled in heat engines. In many cases, it is only secondary, often giving precedence to safety, to strength, to the durability of the engine, to the small space it must occupy, to cost of installation, etc..."

Although Carnot's precise motives were unclear, they most certainly did not pertain specifically to automobile engines. Yet, this timeless statement is both insightful and realistic in describing the challenges of developing powerplants throughout the history of the automobile. His words remind us that the powerplant is not an end in itself, but a result of many, often conflicting factors of which efficiency is but one. The journey, so to speak, is not one of linear logic methodically moving toward a solution, but an often disjointed process of confronting literally thousands of issues and then engineering practical solutions. What follows is an attempt to capture the essence of this paradoxical journey without necessarily including every limb on the family tree of today's powerplants. Several loosely connected vignettes are offered relating to how it was at the start of the journey and how some of the human, technical, and scientific events evolved into the volume-production powerplants of today.

Engine Configuration—How Many Is Enough?

"Why have an eight when you can go 60 mi/h with either a six- or four-cylinder motor?" This question was presented by Mr. John O. Heinze and was hotly debated at a 1915 meeting of the Indiana Section of the Society of Automobile Engineers (SAE). With slight modifications to the technical arguments, similar questions are being debated today—perhaps in an environment confused by the ensuing history rather than clarified by it. The early progression of engine types and number of cylinders was quite straightforward, as Mr. Heinze noted so eloquently: "As you well know, we started at the beginning with a single cylinder motor. We had a great deal of trouble....The next step was to make a two-cylinder motor, which proved more successful." What he failed to note was that, even at the time of his meeting, evolution of the automobile engine was proceeding in a most diverse manner, a far cry from the first logical step from one to two cylinders. To wit, the first eight-cylinder car was apparently produced a year before the first six-cylinder car, and the first high-volume car with a five-cylinder engine would not come along until 1974. Cylinder arrangement and combustion-chamber design are key elements to provide insights into the intensive, and often circuitous, development that has resulted in today's automotive powerplants.

Pre-automotive engines had simple combustion chambers, essentially a flat plate closing the cylinder, and were aspirated by slide-type valves—a hold-over from prior steam-engine experience. As technology evolved from atmospheric cycles to compression of the charge before combustion, valves capable of sealing ever higher pressures and managing more heat were required. Poppet valves, successfully used on pumps for many years prior to gasoline engines, were an obvious choice for early designers. The first motorcycle engine by Gottlieb Daimler and Wilhelm Maybach in 1885 had a suction-operated inlet valve, an exhaust valve located directly under the inlet valve, and an unusual third valve in the piston; all three were poppet type. The combustion chamber could loosely be described as hemispherical, although the valves were located in an adjoining volume.

The first two decades of automotive engineering produced dramatic improvements in combustion, ignition, and volumetric efficiency with specific output rising from approximately 1.5 kW/L in 1896 to an average of approximately 7.5 kW/L in 1914. Even so, engine designers of this period were forced to meet customers' ever increasing desire for performance by creating engines with enormous displacements. Engines as large as 11.5 L were not uncommon in large road

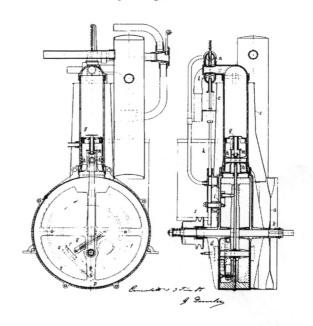

Daimler and Maybach engine for the first motorcycle, 1885. (Source: Internal Fire *by Lyle Cummins.)*

cars, with racing models as large as 26.2 L! Starting in 1902 when Locomobile introduced the first water-cooled inline four-cylinder engine in the United States, four cylinders quickly became the configuration of choice. After 1915, the manufacturing genius of Henry Ford and his Model T would ensure the dominance of the four, based on sales. By 1906, six cylinders were in limited models such as Pierce-Arrow, Stevens-Duryea, and Franklin, with more to follow: Oldsmobile in 1908, Buick in 1910, and Packard in 1912. A few eights were produced during this period, most notably a V8 produced by French automakers De Dion and

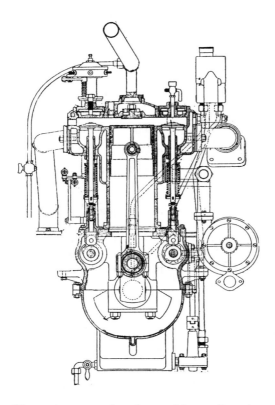

Pierce-Arrow produced one of the earliest six-cylinder engines. The end section is shown. (Source: SAE Bulletin, Volume II, April to September 1912.)

Bouton. A survey of the leading 1914 American models showed one two-cylinder, 54 fours, 45 sixes, and no eights; average displacement was 5.7 L, and average output was 25 kW.

The alphabet soup of "T," "F," "L," and "I" cylinder heads appeared in a seemingly unpredictable manner as designers struggled to improve output. Starting with Mercedes in 1901, the T-head, with intakes on one side of the bore and exhausts on the other, became quite popular, but by 1910 were being displaced by less expensive L-head designs. By 1917, L-heads accounted for approximately 70% of the engines produced, but overhead valve designs (then called valve-in-head) were becoming popular, used in approximately 20%. The Buick Light Six was a typical production overhead valve unit of the day; it featured an I-head, roller lifters, and an exposed pushrod valvetrain. Four valves per cylinder had captured the eyes of race-engine designers, an example being the Wisconsin-Stutz 4.9-L racing motor; this sixteen-valve inline four developed 97.6 kW at 3000 rpm.

Regardless of configuration, all the pre-World War I poppet-valve designs were noisy, unreliable, and required frequent adjustment. These shortcomings kept some designers and numerous inventors from abandoning

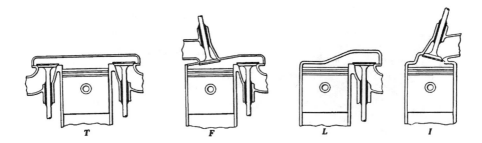

The alphabet soup of cylinder heads.

sleeve-valve technology. Of the many efforts, the most notable design was the double sleeve valve invented by Charles Y. Knight of Indiana. He was clearly focused on quietness of operation, referring to his engine as the "Silent Knight" during the 1906 Chicago Auto Show. Knight engines achieved remarkable early success in England, where, by 1910, Daimler-Knights accounted for nearly a quarter of British auto production. Knight received patents in eight countries, and his engines were ultimately produced by 30 firms. In the United States, Knight had a more difficult time. Rejected by Packard, Locomobile, and Peerless, he finally licensed F.B. Stearns and Willys Company in 1911, and these became the main U.S. producers of the sleeve valve. By April 1932, when key patents

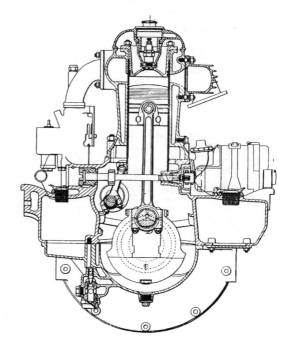

Sectional view of Willys-Knight sleeve valve engine. (Source: SAE 1926 Transactions, Part 1.)

Cadillac V8s introduced in 1915 (top), 1928 (center), and 1936 (bottom). (Source: SAE Quarterly Transactions, 1949, Volume 3.)

expired, improvements in poppet-valve engines had made Knight sleeve valves comparatively heavy, poor in oil economy, and expensive. Production ceased shortly thereafter, but its legacy is one of having intensified development for quietness in engines of all types.

With large-displacement inline fours, vibration was a major problem, and it was not long before market demand arose for more cylinders. Alhough not the first in production worldwide, Cadillac led the headlong rush to V8s by introducing the first mass-produced version in 1914. The 5.1-L 90° L-head design featured two integral block and head castings in iron mounted on an aluminum crankcase. The original version had a single plane crankshaft that had unbalanced, second-order shaking forces and undoubtedly failed to deliver the smoothness of operation demanded by ever more sophisticated customers; it was redesigned in 1923 with counterweighting and a two-plane crank. Peerless followed in 1915, and by 1916 there were eighteen companies building V8s. Low-priced cars would not see V8s until 1932 when Ford introduced the first of the legendary "Flatheads," setting a benchmark for simplicity and cost effectiveness. Somewhat surprisingly, straight eights did not come along until considerably later, with the first mass-produced version introduced by Packard in 1923. By 1930, it had become the most prevalent arrangement in premium vehicles including Auburn, Duesenberg, Locomobile, Marmon, Pierce-Arrow,

Cadillac V8 of 1915.

Studebaker, Chrysler Imperial, and Stutz. Straight eights endured in the United States until Packard ceased production in 1955.

Eight was clearly not the only answer for engine designers searching for luxury and refinement. In 1915, Packard debuted the famous "Twin-Six," a 60° bank angle 6.9-L V12 with such notable refinements as aluminum pistons and roller tappets. Smoothness of operation was clearly the overriding design consideration driving ever increasing numbers of cylinders, as engine mounting technology of the time was crude and engine vibrations were transmitted directly into the bodywork. Twelve would not prove to be the end, with Cadillac introducing a V16 in 1930 and Marmon following in 1931. The Cadillac unit was a 45° bank design at 7.4 L with an overhead valve arrangement that included roller followers and hydraulic lash adjusters. Cadillac and Marmon were the only manufacturers to produce sixteen cylinders in volume.

Shortly after World War I ended, two tremendous advances occurred in the understanding of combustion and, hence, combustion chamber design: the development of tetraethyl lead as a practical anti-knock additive and Sir Harry Ricardo's "Turbulent Head." Geometrically speaking, Ricardo's new version of the L-head chamber had the clearance volume displaced toward the valves. This design reduced knock by producing turbulence via compression-induced lateral flows, today known as "squish." Ricardo's

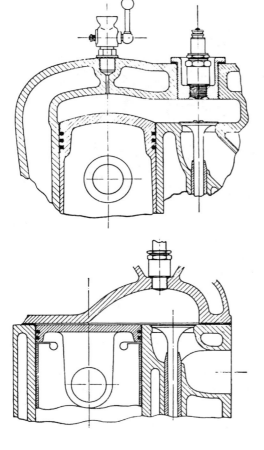

Two L-head types: an "old style" (top) and the Ricardo "Turbulent" (bottom). (Source: SAE Paper 760605.)

work came swiftly to the United States, via a 1921 SAE paper, and attracted the attention of engineers at General Motors, Chrysler, and Packard. The Chrysler 1924 "High Compression" six was an immediate result. Packard and Pontiac designs soon followed, and the turbulent L-head became the accepted design standard for two decades.

The turbulent head was more than a specific design; it embodied a new understanding within the industry of the impact of combustion chamber mechanical design on knock-limited performance. As an example, one of the more subtle results was the idea of locating the spark plug in a hot and compact region of the combustion chamber to burn first the end-gas most likely to undergo self-ignition and produce knock. This line of thinking carried across to OHV engines as well, as evidenced by Chevrolet inline sixes from 1934 to 1961 which used compact, "blue flame" chambers with the spark plug placed in an exhaust valve pocket and a vertical intake located in a large "squish" area.

As World War II ended, the dominance of the L-head was apparent. Cadillac, Pontiac, Oldsmobile, Ford, Chrysler, Studebaker, and Hudson were all committed to L-head designs. However, intensive research at

General Motors was about to spark yet another rebirth of creativity in engine design. In a 1947 SAE paper, Charles Kettering presented the somewhat shocking conclusion that compression ratio could essentially double from established practice to 12.5:1 with high-octane fuel. Cadillac did not wait long to put research into practice with production of the 1949 V8. Early development on this project had clearly shown the disadvantages of using L-heads at high compression ratios, and an overhead-valve combustion chamber capable of a compression ratio as high as 12:1 was selected (the production version was 8:1). Oldsmobile followed the next year with a similar design for the "Rocket V8." Shortly thereafter, Buick, Pontiac, and Chevrolet released similar designs—as did Ford. The now-familiar wedge-shaped chamber cross section was to become a standard practice for decades, but not everyone in the industry followed General Motors' lead on cylinder-head design. Engine designers at Chrysler

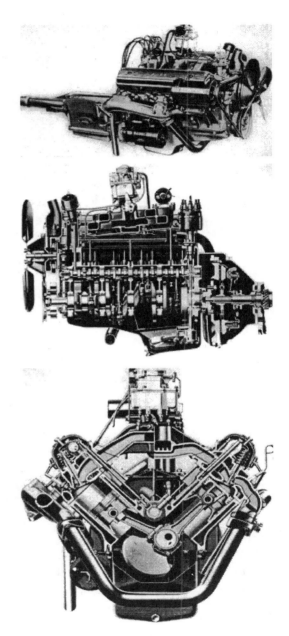

1949 Cadillac V8. (Source: SAE Quarterly Transactions, 1949, Volume 3.)

focused more on volumetric efficiency and the effect of a central spark plug on octane requirement. Their work led to an entirely different solution, the "Hemi" Head introduced in the 1951 Chrysler 5.4-L V8. By 1953, V8 production exceeded six cylinders, and the horsepower race was on!

For more than two decades after the Cadillac V8, designers in the United States painstakingly refined the overhead valve arrangement. Abundant fuel supplies, low fuel prices, and generally good economic times again resulted in displacement, rather than engine technology, as the

Chrysler 5.4-L "Hemi" Head V8 valve gear, 1951. (Source: SAE Quarterly Transactions, 1951, Volume 5.)

preferred means to increase output. V8 displacement increased progressively to a maximum of 8.2 L as designers concentrated on stretching existing designs and cost reductions. Even Chrysler was forced to drop the "Hemi" Head in favor of cost-effective, lightweight wedge chambers. On the other end of the product spectrum, rising import sales in the United States resulted in "compact" cars of the early 1960s and renewed interest in six cylinders, of both inline and 90° vee configurations. The 90° V6 was a drastic departure from past practice that was heavily influenced by tooling and design commonality with existing V8 engines, reducing the cost of introducing a new engine.

By the late 1960s, emission control requirements had started yet another resurgence of cylinder-head experimentation, as engineers searched for "clean" combustion chamber designs. It ultimately became apparent that aftertreatment, not engine design, would be the primary direction to achieve ever more stringent emission requirements. Only modest changes to combustion chamber designs were ultimately made to production engines in this period, with reduced compression ratios which helped lower oxides of nitrogen and hydrocarbon emissions.

The 1970s brought oil embargoes, fuel shortages, and CAFE (Corporate Average Fuel Economy) legislation to the United States. The dominance of the "modern V8" was coming to an end. The sudden shift in demand for fuel economy sent American designers scrambling to their own research labs and to their counterparts in Europe and Japan for fuel-efficient design concepts. Heron Chambers (bowl in piston), "bathtub" chambers, Ford's "PROCO" stratified charge engine, Honda's "CVCC" prechambers, diesels, and numerous other ideas surfaced to meet the seemingly mutually exclusive requirements for high fuel economy and low emissions. At the same time, vehicle packaging requirements for new, smaller cars were forcing engine designers to rethink configuration again. Reduced displacement and transverse front-wheel-drive configurations drove a dramatic rebirth of inline four-cylinder engines. Aluminum heads with single overhead cam, two valves, and "compact" combustion chambers emerged to become a new standard for cylinder heads, and a new engine configuration, the 60° V6, entered the automotive arena. Why 60 degrees? This quote from a 1979 SAE paper describing Chevrolet's 2.8-L V6 engine succinctly summarizes the answer: "Although there are other reasons for a 60° bank angle, the predominant factor was space."

Concern over fuel availability faded in the 1980s, and, as had happened many times in the past, low fuel prices and good times created demand for more performance. But unlike the past, increased displacement was not the only answer. Fuel-economy requirements, both legislated and market driven, as well as newly tooled carlines with tight engine compartments, placed a premium on technologies that produced high specific output. Four-valves-per-cylinder designs, once limited to race engines or exotic supercars, were now being developed for high-volume engines of the 1990s. Four cylinders were first, followed by optional V6s, and finally even V8 "luxury car" engines were designed with four valves.

Today in the U.S. car market, V6s are clearly predominant, with overhead cam, overhead valve, and two- as well as four-valves-per-cylinder designs spread throughout high-volume models. The lower end of the market still chooses the inline four, and the luxury end has apparently come full circle again to embrace the V8. History suggests, however, that it would be foolhardy to presume that we have arrived at the limit of designers' imaginations. However, strangely enough, some of us find ourselves in 1995 again asking Mr. Heinze's question of 1915: "Why have an eight when you can have a six...?"

Powerplant Perspectives: Part II

Gordon L. Rinschler and Tom Asmus
Chrysler Corporation

Harmonization—The Auto/Oil Struggle for Balance

"...it would be possible to double, treble, or even quadruple the mileage obtained from engine fuel..." "It is only regrettable that he [the previous speaker] does not go further and tell us how to accomplish this." "The burden, therefore, falls upon the automotive engine, which must consequently so adapt itself as to gain higher thermal efficiency..." No, this is not the PNGV (Supercar) debate of 1993, but a 1919 discussion at the annual SAE meeting in New York at which oil and auto industry engineers considered options for confronting imminent gasoline supply concerns. Many such discussions have characterized the often diverse interests between oil producers and car manufacturers throughout the history of the automobile. Behind the occasionally heated rhetoric lies a history of intensive research and remarkable progress that has cycled from discord to harmony between fuel and the machinery that releases and transforms its energy.

At the onset of the automotive era, the fuel of choice was "stove" gasoline, having an end point around 94°C (201°F), and early auto producers crafted fuel systems for compatibility with this fuel. The early proliferation of automobiles increased demand of this fuel to such an extent that fuel producers had to make process changes to keep pace. As fractional distillation of crude was the only refining process of the pre-1913 period, and as crude was in short supply, the only practical option was to broaden the gasoline cut at the heavy end.

By the mid-teens, 177°C (351°F) end-point fuels were common, and the stage was set for tempestuous relationships between auto and oil producers as well as their mutual customers. Volatility reduction produced problems of starting and reliability as spark plug fouling and oil dilution yielded predictable results, particularly in cold weather. Engine designs began to include intake air and manifold heating features in response to prevailing fuel trends. Fuel quality typically was based on volatility alone, and the term "high test" conveyed the idea of high volatility and was often gauged by putting a wetted finger into the wind. With the advent of the thermal cracking process in 1913, more gasoline yield from crude was to follow, but this did not reverse the volatility trend. Underestimation of global petroleum supplies during this period consistently produced uncertainty as to the future of the automobile.

Discord between auto and fuel industries, initially based on volatility, evolved into one based on the anti-knock quality of fuels. The auto industry knew all too well the importance of compression ratio for fuel economy and performance, but available fuels would not allow knock-free operation with compression ratios exceeding 4:1. By 1916 it became increasingly clear that a forum for inter-industry coordination was desirable, and by 1920 Charles Kettering had taken steps to formalize this process. Hence, the first organized auto/oil debates were underway. The hitherto history of the automobile and the advent of aviation exercises of World War I served to focus attention on raising the anti-knock quality of gasoline.

Efforts, mainly within the GM Research Laboratories, led to important deductions regarding the role of fuel chemistry as a factor in knock. Pioneering work by Kettering, Midgley, and Boyd led to a proper understanding of knock and the means to control it with fuel additives. In the period following 1918, several classes of fuel additives were identified which demonstrated anti-knock qualities—among these were iodine and aniline. Several indicator systems crafted from commonly available materials in the aforementioned laboratory led to the significant distinction between pre-ignition and knock and additionally provided metrics for the identification of anti-knock properties of the agents to be tested. Many solubilized elements in the periodic table were tested somewhat randomly at first and later by an ingenious periodic system. Many of these compounds were considerably more effective than aniline but produced foul exhaust odors.

By late 1921, less than a year from the time that program cancellation seemed imminent, tetraethyl lead (TEL) was identified to be approximately 50 times more effective than aniline and did not produce foul odors. This landmark discovery, amid controversy related to health effects, entered the market as Ethyl gasoline. The vision of the mid-1920s held that global petroleum supplies would last only until approximately 1945, whereupon a renewable fuel such as alcohol could carry us into the future. Hence, Ethyl was seen only as a "bridging" fuel or a petroleum extender.

TEL usage began in the late 1920s—at first with small dosages. As TEL became readily available in gasolines, engine manufacturers were quick to respond with compression-ratio increases. This remarkable fuel improvement led to the establishment of the new industrial enterprises based on fuel additives and heightened the need for auto/oil industry cooperation. The need for standardization of methods led to auto/oil joint ventures first in 1932 and again in 1934 on "dyno" hill near Uniontown, Pennsylvania, where road, motor, and research octane metrics were correlated. The motoring public responded enthusiastically to new-found vehicle performance and efficiency, and the trend to higher compression ratios and higher octane fuels continued into the early 1940s.

In the period following World War II, TEL dosages in available gasolines began to escalate, and a new type of fuel-induced engine problem arose. A glaze-like material composed of mixtures of oxides and halides of lead coated exhaust valve and seat surfaces which, if allowed to become excessive, could chip off—leaving a leak channel that would "burn" valves. There was no doubt that TEL was the cause, but fuel changes were not the solution. Engine design changes had to be developed to deal with control of valve-seat deposits. These changes came in various forms including positive valve rotators and flexible valve-head designs that produce a sliding action, wearing the glaze at a controlled rate. Slight dissimilarity in the valve and seat angles was another technique used. Such practices were continued until lead phase-down was mandated in the early 1970s, and those features which had rendered engines lead tolerant now produced excessive valve seat wear when operated on unleaded fuel. Again, resolution would be in changes to engine design, not the fuel.

By 1975, the catalytic converter was widely introduced and unleaded fuel was readily available. Unleaded fuels of this period typically had reduced anti-knock quality compared to earlier leaded fuels; hence, compression ratios were reduced on many models. The coincidence of regulatory pressures and shortages initiated by the 1973 oil embargo produced a condition in which unleaded gasoline became a generic, bulk commodity largely devoid of brand distinctions, and many aspects of fuel quality suffered. Customer and auto industry pressure on the oil industry to raise octane quality of unleaded gasoline produced yet another engine-fuel discordance which manifested itself in the early 1980s as a front-end (Reid) vapor pressure rise. As a direct result, hot fuel handling and driveability problems plagued the mostly carburetted U.S. fleet; carburetor fuel foaming (or spewing) and fuel delivery system vapor lock were commonplace. While the auto industry crafted near-term fixes, the cost and complexity of these already heavily "band-aided" fuel systems rose again. This outcome undoubtedly hastened the demise of the automotive carburetor in favor of fuel injection.

As multi-point fuel injection became dominant following a brief single-point fuel injection stint, a new set of auto/oil issues arose based on fuel-related deposits. Fuel additives, effective in keeping carburetors clean, did not provide adequate cleansing action for fuel-injector nozzles. Furthermore, intake valve deposits that might have been inconsequential with the crude, transient fuel control of carburetors, caused driveability problems with the more precise multi-point fuel injection strategies essential to lowering emission levels. Fuel additives in the form of detergents ultimately provided the necessary relief. Another engine-fuel discord arose as engine designers implemented combustion-chamber geometries having tight clearances between piston crown and cylinder head, i.e., tight squish or bump clearance. Carbonaceous deposit accumulation on these surfaces can produce mechanical interferences and a so-called "carbon knock" during engine warm-up as component thermal expansion occurs at variable rates. Again, detergency appears to alleviate this condition.

Driven by resource, marketplace, and regulatory pressures, there will likely be engine-fuel discord in the future. The mechanisms by which these issues become resolved have been increasingly formalized over time. The most recent step in this process came in 1989 when an auto/oil consortium was formed between these two great industries to undertake a massive emissions-related task. Undoubtedly, the working relationships that evolved between the two industries will positively impact the mechanism by which future problems will be resolved and relegate the heated rhetoric of the past to the history books.

Alternative Powerplants— What Does the Future Hold?

Thomas A. Edison, in an 1895 interview, was asked to comment on the future of automotive propulsion. His response: "As it looks at present, it would seem more likely that they will run by a gasoline or naphtha motor of some kind." He did hedge a bit: "It is quite possible, however, that an electrical storage battery will be discovered which will prove more economical..." For those of us familiar with today's debate surrounding so-called Zero-Emission Vehicles, Edison's prophetic words have a familiar ring, but in 1895 dominance of the gasoline engine was far from certain.

Although great strides in efficiency and specific output were being made worldwide with four-cycle engines, the unreliability of hot-tube ignition and uncertainty of the fuel supply clouded the gasoline engine's future. By the turn of the century, the outcome remained quite uncertain. Of the 2,370 automobiles in New York, Boston, and Chicago combined, 1,170 were steam, 800 were electric, and only 400 were gasoline. Ultimately there would be more than 100 different makes of electric cars and more than 125 different steamers manufactured in the United States alone.

The first automobile to exceed 60 mi/h (97 km/h) was an electric driven by Camille Jenatzy, setting the World's Land Speed Record at 65.79 mi/h (105.88 km/h) in April 1899. While speed was never a particular strength of electrics, they offered advantages of quietness and controllability that early gas engines could not deliver. A most desirable feature was not requiring cranking to start, a particular attraction to women. Despite the efforts of Edison and others, battery technology could not overcome the problem of limited range (typically 25 to 30 mi, or 40 to 48 km) and short life due to repetitive deep cycling. There was at least one significant attempt at a hybrid, the Woods Gas-Electric Car. The Woods concept was remarkably similar to some modern proposals as it included a small gas engine and an electric motor/generator capable of regenerative braking. The 1924 National Automobile Show foreshadowed the end; for the first time, there were no electric vehicles to be seen.

Steam cars proved to be a more formidable contender as an automotive powerplant, with high speed and rapid acceleration clearly demonstrated as superior to early gas and electrics. Years of experience with stationary steam engines led to rapid development of reliable and powerful engines for cars, as early speed records attest. Fred Marriot's boat-shaped Stanley Steamer clocked an official 127.56 mi/h (205 km/h) in January 1906 and hit an unofficial 150 mi/h (241 km/h) a year later. Presidential endorsement of steamers came in 1906 when Theodore Roosevelt chose a 22.4 kW White Model G Steam car for official functions. Long start-up times, poor fuel economy, and limited range before taking on water were major obstacles to overcome, overshadowing the advantages of boundless torque from rest, flexibility, simplified operation with no clutch or gear shift, and, of course, quietness. With nearly 180 manufacturers worldwide, improvements came rapidly to address shortcomings, as exemplified by the efforts of Abner Doble. His last car, the 1930 Model F, employed spark ignition (eliminating the need for a gasoline-fired pilot light), flash boilers for quick start-up, and, most significantly, a condenser to address limited range between water stops. Although a magnificent machine capable of 95 mi/h (153 km/h), the Doble Model F was heavy and expensive; only five or six were sold. The customers had voted.

Despite numerous major development programs since then, no commercially viable alternatives to internal combustion piston engines have entered the U.S. marketplace since the last steamer was sold. Massive industry efforts such as gas turbine programs by Chrysler, General Motors, and Ford, and numerous government/industry cooperative programs on concepts ranging from simple hybrids to Stirling cycle adaptations, have yet to yield production models. Only Mazda's persistent refinement of the Wankel rotary resulted in any significant volume in the automobile market, but the rotary never became a serious challenger to the venerable piston engine. Electrics, it can be argued, are making a comeback attempt today, and the impact of unprecedented government/industry cooperation focused on powerplant research of numerous types has yet to be felt in the marketplace. The next 100 years of automotive powerplants will undoubtedly be as unpredictable as the first, but Edison might have had a point after all!

Pressure Charging—An Interesting Side Trip

Without much doubt, superchargers and turbochargers have been the most exotic, exciting, and controversial devices ever added to automobile engines. This quote from a 1920 reference rings vaguely familiar to very recent history: "The present day tendency toward the use of many valves...would seem to lead back to the subject...of the need for forced induction. This will introduce a greater quantity of gas into the cylinders without resorting to the complications and trouble-breeding possibilities of four valves per cylinder." The evolution of forced induction is deeply intertwined with the development of engines, automobiles, aircraft, and of course, racing, and is exemplary of the complexities of the development process and the many external and technical factors that shaped the products of today.

In early engines created by Daimler, Brayton, Dugald Clerk, and others, some form of pressure charging appears to be inherent in the design concepts. Gottlieb Daimler, in fact, received patents in 1885 and 1889 specifically describing four-stroke engines with devices to pressurize the intake charge; it does not appear, however, that these concepts led to any practical application directly. As the century turned and the auto industry started its explosion of creativity, piston-type superchargers continued to entice inventors but never seemed to progress beyond the prototype stage, probably because reliability was such a dominant focus of the early engine engineers. There are a few documented examples of concepts that clearly demonstrated improved performance, but did not appeal to potential investors and thus never made their way into production. An "air augmentation" system used in a three-cylinder Dawson Car in England was a novel design, and there were also efforts by such notables as Englishman Dugald

Clerk and Frenchman Louis Renault. Then, as in many times since, the accepted way of increasing output was increased displacement, not technology.

Piston compressors soon gave way to rotary designs, typically either vane or Roots types. Ultimately, racing (and its role as both test track and advertising venue) set the stage to bring supercharging from the laboratory to the street.

In 1907, Lee S. Chadwick set about to demonstrate the worth of his new luxury car by racing it in a stock form, albeit stripped of unnecessary body work. The Great Chadwick Six, a huge 11.6-L T-head inline six, was modified with three carburetors for improved performance. When preliminary tests resulted in disappointing speeds, Chadwick and his team hastily designed and built a single-stage centrifugal "blower" that clearly showed the benefits of pressure charging. A maxim that has guided many decisions in engine design, "if a little is good, more is better," led to the final three-stage design driven by a leather belt at an incredible 9:1 ratio; belt slippage at high speed apparently was an inelegant but effective method of preventing compressor overspeed! Chadwick's racing success was phenomenal, wiping out all comers, until an apparent sabotage caused him to drop out of the 1908 Vanderbuilt Cup race with mechanical difficulties. His success did little to promote the concept of supercharging, as Chadwick kept his "speed secret" concealed, presumably to give the impression that the standard car possessed the same high-speed capabilities of the racing version! Ahead of his time, Chadwick's inventiveness did little to benefit engine designers of the day, or himself, and Chadwick ceased production after building only 235 cars in eleven years of production.

Outside racing circles, pressure-charging development was, for the most part, abandoned by automakers but became a major focus of aviation engineers seeking to provide high performance at altitude. World War I and the attendant realization of the military value of aircraft provided added incentive for high-altitude power available with mechanically driven superchargers. In the United States, the same rationale resulted in preliminary development of turbocharging, using energy from the exhaust to drive a compressor. A most notable event was testing a crudely turbocharged

Liberty V12 engine at the summit of Pike's Peak, achieving roughly the same power output at 14,000 feet as at sea level. Although racing efforts continued, peacetime brought an abrupt drop in interest in all forms of inlet pressure boosting, and production supercharging efforts were limited to a few luxury German cars. Significant during this period was the first use of a Roots-type blower and, coincidentally, the hastened demise of the sleeve valve engine when weaknesses in exhaust sleeve lubrication and cylinder heat transfer were exacerbated by Daimler's early attempts to supercharge a Mercedes-Knight.

Starting with the 1924 Indianapolis 500, supercharging arrived for good in racing when Sanford Moss of General Electric adapted a mechanically driven centrifugal blower to Fred Duesenberg's car. The ensuing victory for Duesenberg inspired Harry Miller, a long-time Duesenberg rival, to begin development in earnest. The result was the "Millercharger," claimed by Miller to operate at 37,500 rpm and increase race-engine output from 90 to 151 kW. Duesenberg did ultimately build production cars with superchargers but, despite impressive racing credentials, they were not well received in the marketplace.

After peaking in World War II, military aviation research and development of piston engines ended as efforts were redirected to jet engines. Automakers, focused on retooling to meet post-war demand, had little need for innovation to sell cars and virtually no interest in pressure charging. In the 1950s, supercharging appeared again in production models—the Studebaker Golden Hawk and Ford Thunderbird. Supercharging was far from a marketplace success and was followed by limited application of turbocharging in the 1960s, the Corvair Turbo Spyder and the Olds Jetfire being notable examples.

Turbocharged cars, while always exciting, did not achieve high-volume usage until the 1980s when first Ford, then Chrysler and General Motors, introduced them in production models. The Chrysler family of turbocharged four-cylinder engines was ultimately used in many car lines, even minivans. It was a brief success, as market demand for naturally aspirated V6 engines swept turbos from the showroom floor by the end of the decade. Today, the viability of pressure charging for production engines is again a subject for debate. Supercharged V6 engines have been introduced by Ford and GM but, arguably, these could be marketplace surrogates for V8 engines not yet available in the quantities customers demand; an analog of the 1980s' turbo demise. Meanwhile, those allegedly "trouble breeding" four-valves-per-cylinder engines continue to account for a large portion of the market.

Chapter 3

Powerplant Perspectives: Part III

Gordon L. Rinschler and Tom Asmus
Chrysler Corporation

Two and Other Cycles—This Is How It Started

It can be argued that the last "breakthrough" in internal combustion engine technology occurred in Germany in 1876 when Nikolaus Otto introduced the idea of a four-stroke cycle, and the massive change that has occurred since that time resulted from incremental improvement! Without debating the merits of this argument, it is clear that the four-stroke-cycle concept actually challenged engine-design thinking of its day, contrary to some popular wisdom that the dominance of the four-stroke cycle was a direct result of simply having started there. In fact, Otto's concept was far from easily accepted and survived numerous challenges in the highly competitive market for engines that existed prior to development of the automobile.

Prior to 1876, all internal combustion engines had fewer than four strokes per cycle, and most were atmospheric (noncompression)—not surprising, because the steam engine was the archetype for engine thinking. These had double-acting pistons (a one-stroke cycle), and combustion occurred at atmospheric pressure. The mechanisms to execute these cycles were highly diverse in all respects except for the piston/cylinder assembly. These designs reflect consciousness of the day regarding issues such as the power of vacuum, rate and reliability of cycle replication, the explosive nature of internal combustion giving rise to piston shock, and (for want of a better term) the "waste" of piston strokes.

A popular design of 1867, the Otto and Langen engine, reflects all of these. The vertically disposed cylinder has a piston with an attached rack which engages a gear on a flywheel shaft through an overrunning clutch. Internal combustion drives the piston rack assembly upward, imparting momentum

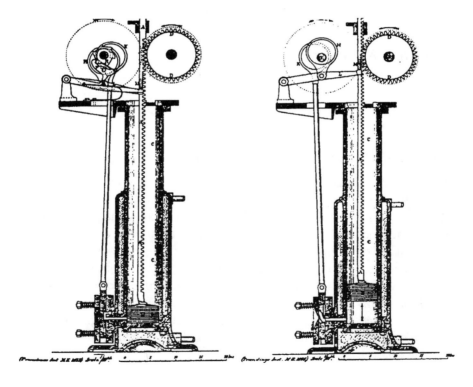

Section view of Crossley's Otto and Langen engine.
(Source: Internal Fire by Lyle Commins.)

to it while overrunning the shaft. This event produces no useful work but overexpands the working fluid (creating a cylinder vacuum) and raises the potential energy of this partially constrained projectile. Downward motion with the shaft gear engaged, "driven" by cylinder vacuum created by over-expansion and by gravity, led to the production of useful work. Given the slow operating speeds employed and the overwhelming problems of reliably creating a combustion event in engines of this vintage, it is easy to understand intellectual resistance to the idea of "wasting" a piston cycle. Otto's four-stroke would have to prove itself.

When four-stroke engines entered commerce, the efficiency and reliability of this engine proved far superior to any preceding design. Deutz, in Cologne, was the original licensee to build this engine under the Otto patent; British and U.S. manufacturers were to follow, producing approximately 50,000 in 1890. Far from stifling creativity, early success of the four-stroke engine encouraged experimentation. Deutz vigorously pursued infringements of the Otto patent, and this precipitated a whirlwind of activity aimed at developing an engine that could compete with the Otto engine without incurring legal problems. In addition to the four-stroke cycle, claims relative to charge stratification were an additional obstacle for Otto's numerous competitors, by precluding minor modifications to prevent patent infringement.

"Cycle" competition appears to have taken on nationalistic overtones with the German Otto, the French Carnot, and the U.S. Brayton. Predictably, several six-stroke cycle designs also emerged during this period. The British could not lay specific claim to a cycle, but nowhere was the two-stroke cycle more vigorously pursued than under the brilliant tutelage of such Englishmen as Dugald Clerk and James Atkinson, whose respective "cycles" were British. Between 1876 and 1890, two-stroke engine development and production were widespread but apparently not widely accepted. An interesting example is the Nash two-stroke engine built in New York beginning in 1888; it is the closest progenitor of the modern outboard marine engine, with crankcase scavenging, two ports, and a reed valve. When the Otto patent finally expired in 1890, the four-stroke engine was accepted as the engine of choice by a competitive market, based on the engine's superior efficiency and reliability.

The accelerated two-stroke activity during the period of the Otto patent protection had many talented contributors, and some lasting ideas were born. The mechanisms conceived for the execution of these two-stroke cycles were highly diverse. Many had separate charging or scavenging cylinders coupled with a power cylinder. The Atkinson engine was of a valveless, opposed-piston design, and its inventor was conscious of the desirability of very rapid expansion (relative to shaft speed) and long expansion stroke (relative to compression). The Atkinson mechanism addressed these thermodynamic ideals of reduced heat loss and increased expansion-work transfer, respectively. Engine designers since that time have pondered mechanisms for the realization of increased expansion-work transfer, although the practicality of most have been challenged on the basis of added friction and weight. More recently, supercharging devices, while generally associated with high performance, may also contribute a small measure of cycle compounding by producing positive pumping work during the induction stroke at some operating conditions. Another more subtle form of cycle modification has resulted from modified valve-timing schemes in conventional automotive engines. It is common for the effective expansion ratios to exceed that of compression, and this is a natural consequence of the optimization process which factors in the control of engine breathing. The much-discussed "Miller" cycle embodies these two aforementioned ideas.

Today, two-cycle development is again proceeding in earnest. From Detroit to Australia, extensive research on fuel preparation and combustion has resulted in remarkable progress on emissions levels, a major obstacle for automotive applications. The marine engine industry is beginning to turn the research into practice as it prepares to meet emissions challenges for two-stroke recreational engines. All that can be stated with any certainty is that, despite numerous challenges, four-cycle engines have dominated the history of automobiles, and the future of the two or any other cycle operation for cars will be determined, as in the past, by demonstrated performance. Perhaps there will be another "breakthrough."

Exhaust Emissions—Forcing Technology, A New Dimension

In the opening line of his comments on a 1913 paper on carburetor testing, Mr. E.R. Hewitt stated, "I began work on exhaust analysis in about 1898, and I have been following it through pretty thoroughly ever since." While this may not be a "first" or even a significant historical event, the quotation demonstrates that exhaust gas analysis, and hence the study of engine emissions, is as old as the automotive powerplant itself. The difference between now and then is one of focus. The early work used measurement of certain exhaust emissions as a development tool for engines and fuel systems, whereas the modern focus is on the emissions themselves and controlling them as a primary design requirement.

Early work on combustion addressed the most basic concern—did it occur? Providing a combustible fuel/air mixture and an ignition source at the proper time were enabling developments that were essential to the success of the automobile itself. Many pre-1900 engines use external pilot burners either to conduct heat through a "hot-tube" inserted into the combustion chamber, or in conjunction with a "flame" ignition system that transported an external flame into the combustion chamber. Others tried mechanical "make and break" approaches in which a cam would close, then open, an in-cylinder electrical contact. None of these approaches yielded reliable combustion. However, the goal of early engineers was simply to keep the engine running, and thus an occasional misfire was quite acceptable.

By the 1920s, ignition systems had progressed to two relatively reliable approaches: vibrator-based induction coils (passive spark plugs such as those used in the Ford Model T) and single-coil, single-spark, breaker-point distributor systems. Automotive engineers still were concerned with whether or not combustion occurred, but they had moved on to another question—did it knock? Sir Harry Ricardo and his contemporaries found that certain combustion-chamber geometries reduced knock. Ricardo also recognized that fundamental thermal efficiency losses were attributable to increased working-fluid specific heat at flame temperatures during the expansion stroke and direct cylinder-wall heat loss. He attributed 30–40% efficiency loss to these factors and reasoned that "weak" fuel/air mixtures could serve to minimize these losses, but some form of mixture stratification was required. The concept of "lean" burning and related developments to facilitate combustion evolved with fuel economy as prime motivations.

Detailed observations of in-cylinder processes would not be reported until the 1930s with the GM Research spectroscopic and photographic recording of normal and knocking combustion along with pre- and post-combustion chemical phenomena. By the late 1930s, in-cylinder flow studies in motored, glass cylinder "engines" were reported by the Langley Memorial Aeronautical Laboratory. Shrouded and pinned valves in two- and four-valve designs were studied using feathers to seed the flows. Swirl and tumble flows were observed, and the importance of large-scale flows and their persistence during the compression stroke was recognized. Again, these and other studies of the day were directed at improving efficiency and combustion robustness; exhaust chemistry was merely one of the tools used in the engineering laboratory.

The dramatic change in focus began in the 1950s when Professor A.J. Haagen-Smit made the connection between hydrocarbons (HC), oxides of nitrogen (NO_x), and the formation of "smog" producing ozone in the atmosphere. Because gasoline engines produce both chemicals, albeit in small concentrations, the automobile was soon implicated by the public and its elected officials as a major cause of pollution. Solutions were demanded, despite the fact that there was virtually no science available to address how that solution would be structured; the concept of "forcing" powertrain technology through government mandate was born.

The first result was a frenzy of activity in Detroit to understand the mechanisms by which the powerplant produced unburned HC and NO_x. Industry protests that cars were being unfairly burdened with controls—and some remarkably poor attempts at public relations—caused a public perception that the industry was "foot dragging" on emissions controls. However, in the technical community, automotive engineers and researchers were doing a remarkable job in addressing the sources of engine emissions. A working knowledge of the mechanisms by which exhaust emissions occur and the impact of operating and design variables on HC, CO, and NO_x was developed in short order and in parallel with suitable instrumentation and test techniques.

The engine, being the source, was the first logical point of attack for emissions control technology. Crankcase ventilation systems were the first "add-on" emissions controls to appear nationwide in 1963. By 1968, federally mandated control of HC and CO was a reality, and various control devices were added to modify spark timing and fuel/air ratio at various operating conditions. Changes to the base engine included lower compression ratios and combustion chamber modifications to reduce "crevices" and to lower surface-to-volume ratios—both aimed at HC reduction. Deceleration controls, pre-heated induction air, and secondary air pumps were developed for both HC and CO control.

By 1973, NO_x control was required, and exhaust gas recirculation (EGR) was added to the long list of control devices to meet emissions regulations. Combustion research shifted away from efficiency, *per se*, to improving dilution tolerance for more effective use of EGR. Unfortunately, the combination of an emerging, but limited, understanding of the combustion process, and the necessity for rapid development of "add-on" devices to meet government-mandated timetables, resulted in many compromises. Customers were confronted with new cars that had poorer fuel economy and deteriorated driveability.

With few holdouts, by the mid-1970s the industry generally concluded that "cleaning up" emissions at the source alone was not a viable approach to ever-tightening emissions standards, and oxidizing catalytic converters became standard equipment. Although expensive, catalysts shifted some of the burden of emissions control from the engine proper to after-treatment, allowing more robust calibrations for fuel economy/driveability. Durability was an immediate concern because unleaded fuel was required, taking valve and valve seat durability, from a fuel standpoint, back to the 1920s. Hardened valve seats, upgraded valve materials of a stiffened design, and restriction of valve rotation were rapidly introduced to provide wear resistance.

In 1981, federal standards for HC, CO, and NO_x were lowered to a level requiring three-way catalysts with the ability to control all three emissions. Successful operation of these catalyst systems required very precise control of fuel/air ratio and created intense development of feedback fuel-control systems, engine-control electronics, and engine fuel systems. In roughly a decade, carburetors gave way first to single-point fuel injection systems and finally to multi-point fuel injection typical of today's automobile.

Controls and aftertreatment clearly surfaced as the main tools of emissions development engineers, but simply complying with emissions laws was not enough. Customers were demanding better performance, better economy, and flawless driveability. A little-recognized benefit of the intensive efforts on emissions by industry and the scientific community was a windfall of knowledge about the details of chemical, thermal, and fluid-mechanical processes associated with combustion. Engine designers had the opportunity to apply these insights to a wave of new engines driven first by CAFE (Corporate Average Fuel Economy) standards in the early 1980s and then by a rebirth of performance that remains apparent today. Aluminum cylinder-head designs included new combustion chambers and ports with controlled charge motion that allowed higher compression ratios, reversing the downward trend caused by the introduction of 91 RON unleaded fuel. Electronic controls were applied to detonation sensing, so high-performance engines could be developed to benefit from new high-octane unleaded fuels but run satisfactorily on unleaded "regular" gas. After a decade of refinement, powertrain technology had produced cars that could meet prevailing emissions standards without compromising customer demands for smoothness, power, and fuel economy.

Today, however, we are embarking on another dramatic reduction in allowable emissions levels while satisfying new requirements for self-diagnostics, durability, and evaporative control. It appears that again technology-forcing legislation will obsolete "mature" engine technologies and reshape the powertrains to come.

Conclusions—There Are None

Because the goal of Chapters 1 through 3 was perspective, a few summary observations seem appropriate. Throughout the 100 years of the automobile, powerplant technology has literally been "swarmed" by hundreds of incredibly talented engineers and researchers, yet we find a paradox: the piston engine is at one moment a "mature" technology but is regularly radicalized to meet the ever-changing needs of the society it serves. In the early days, one might argue that change was driven by invention, but clearly the preponderance of technological change in automobile engines has been motivated by circumstances in the marketplace or by government policy.

After a hundred or so years, the only conclusion to be drawn is that the world around us will continue to stimulate the creativity and inventiveness of automotive engineers and, as practitioners of the powertrain art, we are on a journey that has no final destination.

U.S. Passenger Car Brake History

Larry M. Rinek
SRI International

Carl W. Cowan

As a subset of the chassis, the braking system plays a critically important role in the safe operation of any passenger vehicle. Brakes on passenger cars have evolved significantly over the past 100 years of the U.S. industry. From crude mechanical devices on the driveline, brakes evolved into cable- or rod-actuated wheel brakes, usually drum type, first for rear wheels, then all wheels. What followed was the addition of hydraulic actuation, some with vacuum power boost, then ultimately caliper/disc types enhanced by electronic controls.

This chapter highlights the key trends, plus side issues such as the evolution of brake friction materials. The main focus here is on service or "foundation" brakes, whose job entails stopping or slowing cars by converting vehicle kinetic energy to frictional heat (dissipated ultimately in the air). Brief mention is also made of mechanical parking brakes.

The Early Years— Mechanical Brakes

Two views of rear external contracting band brakes from a 1902 Rambler Model C Runabout. (Courtesy of Larry Rinek.)

Service brakes were practically an afterthought on America's early production cars. The archives suggest that if designers added brakes, they were most likely on the driveline—such as external clamping of a rotating drum via a band or something similar, lined with friction material (e.g., leather). An example was the 1902 Oldsmobile Model R, with external contracting (band) brakes. Some very early cars also were believed to have used mechanical scuff brakes acting against the rear wheels/tires, a practice borrowed from horse-drawn carriages and wagons.

Ford's original Model A (1903) incorporated a driveline brake in the transmission. This practice continued at Ford with the famous Model T (1908 to 1927). Because road speeds were fairly modest in this early era (10–20 mi/h), the demands on brakes were not large, but that would soon change. Vehicle horsepower and weight, as well as top speeds and roadway safety concerns, increased, leading to superior mechanical wheel brakes such as internal-expanding drum brakes containing twin, semicircular, pivoting brake shoes. The 1910 Sears Model P car literature boasted of "double acting expanding" brakes. Drum service brakes first appeared on the rear wheel (upgraded parking brakes) and then gradually migrated to all four wheels. Ford's trend-setting Model A of 1928 featured four-wheel, mechanical drum brakes. GM's various models transitioned to four-wheel mechanical drums in the 1924 to 1928 time frame (Chevrolet in 1928). Another important technical development in mechanical drum brakes was the self-servo, double leading shoe design (with rod linkage joining shoes) introduced around 1930 (e.g., Bendix Duo-servo on 1931 Lincoln), but this was complex and costly.

As touring and luxury cars grew to ponderous weight levels, power assistance to the mechanical brakes became necessary. Such "free" power boost became possible with the engineers' harnessing of the intake manifold partial vacuum available upon deceleration. The 1932 model Cadillac V16 has 16½-inch (42-cm) diameter drum brakes at each corner of the 6300-lb (2858-kg) car. A vacuum-assist unit controlled by brake pedal travel was on this Cadillac, and final actuation was via flexible cables. For the 1932 model year, Lincoln introduced a

Bendix Brakes advertisement, circa 1930s.
(Courtesy of Northern Indiana Historical Society.)

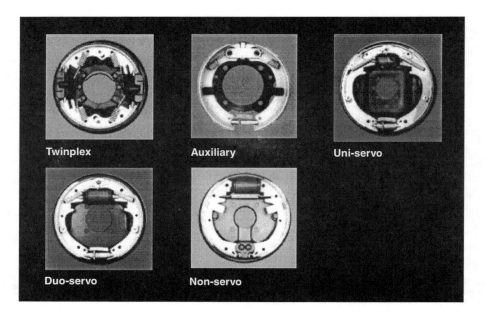

Early Bendix drum brakes. (Courtesy of Northern Indiana Historical Society.)

full-sized V12 car with a 145-inch (368-cm) wheelbase. The four, cable-actuated drum brakes featured a vacuum booster unit on this Lincoln. In 1932, industry pioneer Vincent Bendix (who had been making brakes in South Bend, Indiana, for 10 years) announced the development of his vacuum-boost, power brake system. Besides Bendix, Kelsey-Hayes and Midland Ross later provided vacuum-boost units for cars.

Hydraulic Brakes

The next major innovation was the advent of hydraulic brakes. Now, drum brakes could be applied more forcefully, with better stopping distance and less pedal effort. Use of a relatively large-surface-area, master-cylinder piston, driving fluid fed to smaller-area slave cylinders at each wheel, provided the mechanical advantage.

The first U.S. example of car hydraulic brakes was in the Duesenberg Eight; the prototype introduced to the press at the New York Auto Salon in November 1920 featured four-wheel hydraulic brakes developed by

Lockheed for the aircraft industry but used for the first time in a passenger car. Series production of the Model Eight, later known as the Duesenberg Model A, began in 1921.

The first Chrysler-nameplate, four-wheel hydraulic application appeared in 1924. Information on this new brake technology began to appear in the Society of Automotive Engineers (SAE) literature. GM and Ford were later adapters of hydraulic-brake technology (in the 1934 and 1939 model years, respectively). See Table 4-1 for a collection of U.S. car hydraulic brake introductions. GM and Ford both hesitated to provide hydraulic brakes throughout their lines until servo-type, vacuum-boost units were more proven. GM added power boost in 1935 to 1936. By contrast, Chrysler took more risk with vacuum-boost hydraulics, appearing initially in the 1932 model year Chrysler Imperial and Imperial Custom.

Table 4-1
Hydraulic Brake Releases for Selected U.S. Passenger Cars

Make	Model	Release Year	Remarks
Ford	Lincoln-Zephyr	1938	1939 MY
	Ford	1938	1939 MY, no power boost until 1950s
Chrysler	B-70	1924	Derived from Maxwell
	Imperial	1926	—
General Motors	Cadillac	1934	Model Year
	LaSalle	1934	Model Year
	Oldsmobile	1935	Model Year
	Chevrolet	1936	Model Year
	Pontiac	1936	Model Year
	Buick	1936	Model Year
Packard	—	1935	120 Series
Duesenberg	Eight	1921	667 four-cylinder cars built through 1928 (Models Eight/A)
	J	1928	1929 Model Year*
	SJ	1932	Vacuum servo boost

Table 4-1 (continued)

Make	Model	Release Year	Remarks
Auburn	—	1929	Discontinued after 1932
Cord	L-29	1929	5000 front-wheel-drive cars built through 1932
	810/812	1936	Discontinued after 1937, front-wheel drive
Reo	Flying Cloud	1928	Discontinued after 1932

* The 1929 J started a series of large, 7-L, straight-eight, DOHC, four-valves-per-cylinder cars that totaled approximately 481 production units (Models J, SJ—the supercharged J, SSJ, JN).

(Sources: Carl Cowan, SAE Brake Committee/Passenger Car Activity; James Wagner, SAE Historical Committee; *The Birth of Chrysler Corporation and Its Engineering Legacy*, by Carl Breer, SAE R-144; *Motor Annual*, various years; original car literature; *American Heritage*, July/August 1994.)

Real volume with power-assist hydraulics did not materialize until the 1950s and 1960s. Note the passenger car releases in Table 4-2. By the 1970s and 1980s, power-assist hydraulic brakes became standard on all full-size cars and available for most small cars built in America.

Table 4-2
Vacuum-Boost Power Hydraulic Brake Releases for Selected U.S. Passenger Cars

Make	Nameplate	Release Year
Ford	Ford	1954
GM	Chevrolet	1954
	Corvette	1963
	Buick	1953
Studebaker	Studebaker	1955

(Sources: Carl Cowan, SAE Brake Committee/Passenger Car Activity; James Wagner, SAE Historical Committee.)

Another important hydraulic safety feature announced to manufacturers in 1960 by Bendix (first production was for MY 1962 Cadillac and Rambler) was split-circuit brakes with dual master cylinders. Thus, in the event of a severed hydraulic line, half of the braking power (two front or two rear) remained available. Later, U.S. OEMs adapted the safer diagonal split configuration (one front/one rear), pioneered by Saab in the mid-1960s.

Parking Brakes

Parking brakes originally were intended to hold stopped cars in place, e.g., on a hill. The first wheel brakes (invariably on the rear) were actually parking brakes, later upgraded to service-brake duty. Later, these brakes served a useful safety function in the age of hydraulics as emergency mechanical brakes, should the hydraulic system lose pressure or fail. Mechanical cable systems to the rear brakes were rigged to a driver-operated, ratcheted foot pedal or handle, accompanied by dashboard warning lamps. All automatic transmission cars, starting with the 1948 Dynaflow Buick, included a "park" position on the gear selector to supplement the parking brakes. This mechanical interlock in the driveline proved very effective.

Caliper/Disc Brakes

U.S. passenger cars received a quantum jump in braking performance with the advent of caliper/disc brakes. Proven in racecar applications in the 1950s, some abortive, passenger car, disc brake efforts appeared in the low-volume Chrysler Crown Imperial (1949 to 1955) and Crosley (Hotshot/Super Sports). Volume U.S. series production for passenger cars did not begin until the 1960s. This trend was encouraged by new U.S. regulatory standards for stopping distance, as well as competitive pressures and safety needs related to high-speed cars.

As compared to drum brakes, the caliper/disc brake offered various advantages (although at higher cost per wheel):

- Superior stopping performance (shorter distance)

- High torque development from a more compact and lighter package

- Higher resistance to thermal fade

- Better resistance to water flooding (self-wiping)

- Self-adjustment (no periodic mechanical slack adjustments)

Bendix vacuum booster for 1962 Studebaker, an Avanti with Bendix disc brakes. (Courtesy of Northern Indiana Historical Society.)

Bendix and Kelsey-Hayes were the two U.S. pioneers in proliferation of this technology for passenger cars. Studebaker's futuristic Avanti, announced in the spring of 1962 for model year 1963, "broke the ice" with standard front disc brakes believed to be Bendix-furnished, beginning in late 1961. The 1965 model year saw the first, real, volume production with standard front discs (Kelsey-Hayes) on the Ford Thunderbird and Lincoln Continental, and four-wheel discs on the Chevrolet Corvette. Disc brakes soon became a widespread option, as on the Ford Mustang, but by 1968 they were becoming fairly standard for front brakes, where 60–70% of the braking power was concentrated. By the 1990s, a substantial share of U.S. passenger cars received standard, four-wheel disc brakes. The typical construction of the U.S. caliper/disc system evolved into a ventilated, cast gray iron, disc rotor clamped by a cast, ductile iron caliper containing two opposed pistons. In the 1990s, brake builders were experimenting with lighter and more advanced disc rotors and calipers for passenger cars such as Al-based, metal matrix composite (MMC) types. The SAE literature has had various papers over the years dealing with one of the few disc brake maladies: brake "squeal." The future, however, is assured for disc brakes—the technology of choice for cars.

Brake Friction Material Evolution

Wear surfaces on early brakes did not receive a good deal of engineering effort. Leather, wood, and rubberized or treated woven materials fulfilled the early duty. As speeds of cars increased

Early fixed caliper from ITT Automotive.

and better roads were available in the 1920s, it was necessary to improve the life and performance of brake linings. This required better, more heat-resistant materials. With SAE support, rigorous tests and engineering efforts were put into this area by 1923. Chemists came up with asbestos-fiber-reinforced materials with organic resin binders (thermosets). These were molded, heat cured, and riveted to brake shoes. The recipe for brake linings became proprietary and usually included small quantities of certain "friction modifiers." Asbestos, a noncombustible, fibrous, mineral silicate of calcium and magnesium, provided good wear properties under high-heat conditions. However, due to the (later proven) carcinogenic properties of airborne asbestos fibers, the U.S. government banned the use of the material in brakes beyond 1993. For many years, the industry has been working with asbestos alternatives such as fibers made of glass, metals, rock wool, carbon, Kevlar, ceramics, and other substances/materials. The transition has now been made with success.

Early floating caliper from ITT Automotive.

ITT T52 9-inch vacuum booster.

Another friction material deserves special mention for high-performance passenger cars: the sintered, semi-metallic, disc brake pad (powder-metal-based). With better high-temperature performance than organic matrix materials, the semi-metallics were first proven in racing and then placed on sports cars such as the Chevrolet Corvette. Bendix semi-metallic pads first appeared in 1970. However, this material has a lower coefficient of friction, needs more clamp force, and is hard on disc rotors. For the ultimate in brake performance, carbon/carbon composite disc rotors and friction pads would be the choice (as used in jet aircraft and Formula 1 racecars) but remain a distant dream for production passenger cars because of prohibitive cost.

Electronic Braking Controls

This subsection of brake history mainly concerns the development of electronic, anti-lock (originally known as anti-skid) braking systems (ABS) and the related traction control system (TCS) option. The historical significance of ABS is huge: ABS may represent the single greatest advancement in automotive braking since the development of hydraulic brakes.

ABS has recently become so popular in the United States that it grew from roughly 10% of 1991 MY car production to an estimated 50% (or more) of 1995 production. It has now become a "mainstream" braking feature, standard on many platforms, that offers obvious safety benefits (such as steerability) to drivers on slippery road surfaces.

The rich history on ABS development is briefly summarized in SAE J2246. In essence, the technology was born in the railroad industry in the early 1900s, progressed to cars (in concept) by 1936 with a Bosch patent for electro-hydraulic ABS, and then

Ford Mercury Marquis front disc-brake rotor, 1979 to 1991. (Courtesy of Carl Cowan.)

Ford 10 x 2½ rear drum, 1979 to 1991. (Courtesy of Carl Cowan.)

became a production item for aircraft (late 1940s, early 1950s). U.S. automotive interests experimented with mechanical and analog electronic ABS in the 1950s (Ford, Kelsey-Hayes, Chrysler). By 1968, the first U.S. production installations began: the Kelsey-Hayes "Sure-Track," single-channel, vacuum-powered, rear wheel ABS on Ford's 1969 MY Continental MK III.

The first U.S., four-wheel, electronic (analog) ABS series release is attributed to Bendix on the Chrysler 1971 MY Imperial. This Bendix three-channel, four-wheel, vacuum-actuated ABS was marketed as the "Sure-Brake." These early ABSs, with limited performance, less-than-desirable reliability, and relatively high cost were not popular and soon faded from the scene.

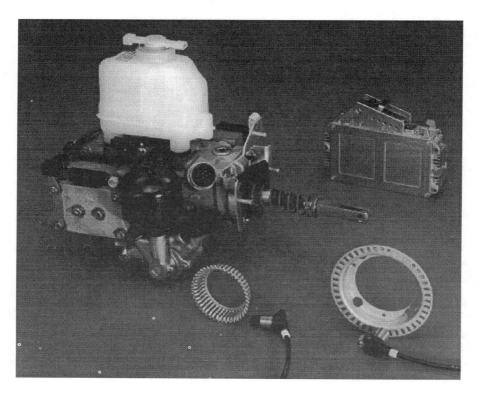

ITT Teves ABS MK II with ECU and sensors (1984 introduction).

It was up to the German engineers to get ABS back into passenger cars in a big way. With Bosch's MY 1979 Mercedes-Benz introduction of a reliable, all-digital, electronic ABS with separate hydraulic booster, ABS was back in business. The U.S. industry became involved again in 1984 with the 1985 MY Continental MK VII featuring a Teves MK II (ITT Automotive) "integrated" ABS with digital microprocessors (dual redundant) and combined master cylinder, hydraulic brake booster, and solenoid-valve, antilock actuators. All of these modern, four-wheel, electronic ABSs rely on wound-coil, variable reluctance sensors at each wheel, mounted next to a rotating toothed ring, that feed sine waves of variable frequency (and amplitude) to the ABS computer, which then calculates wheel velocity and acceleration. Various proprietary algorithms are used to determine whether one or more wheels have slowed too quickly (excess slipping), thereby commanding brake pressure release and reapplication on a pulsating basis.

More advanced ABS generations—e.g., Teves MK IV/MK 20, GM Delco Moraine MK VI, plus equivalents from Bendix, Bosch (ABS 3, ABS 5), and Kelsey-Hayes—are now rolling through the U.S. passenger car fleets with lower OEM cost, higher volume, better performance, and more compact modular packaging. A small share of ABSs are now fitted with extra-cost electronic TCS add-ons (software and minor hardware changes). According to Kelsey-Hayes, only about 5% of North American (production) vehicles currently receive TCS. By modulating service brakes and engine torque (via electronic fuel injection or throttle intervention), more effective acceleration on low-coefficient-of-friction surfaces is feasible with TCS. TCS production first began in Europe (Bosch, Teves), but by 1986, U.S. developers (e.g., Bendix) announced their products. The first U.S. production TCS appeared on the Cadillac Allante (1990 to 1991). This technology is now standard on certain high-end, U.S. production cars (e.g., GM's Chevrolet Corvette and Cadillac and Ford's Lincoln MK VIII since MY 1993). The option is becoming more available on other platforms but is not likely to become as widespread as ABS, particularly for "sunbelt" customers free of ice and snow.

Future Activities

The next big news for braking systems on U.S. cars will again involve advanced electronic controls. More specifically, we refer to the ABS/ASR follow-on variously known as "handling control system," Vehicle

Dynamics Control (Bosch's VDC), Interactive Vehicle Dynamic Control System (AlliedSignal Automotive), Integrated Vehicle Dynamics (ITT Teves' IVD), Electronic Steering Program (Mercedes-Benz), Automotive Stability Management System (ITT Teves' ASMS), Vehicle Stability Management (Kelsey-Hayes), and active directional yaw stabilization. In principle, brakes are modulated (along with engine torque via EFI), aided by wheel speed, lateral acceleration, steering angle, and yaw (or yaw rate) sensors to forestall spinout, without driver intervention. The promise of "no brain" superior vehicle handling for any driver is undoubtedly a big safety breakthrough, although system cost and complexity will initially limit the technology to high-end cars.

Beyond this technology (into the next century), engineers dream of "brake-by-wire" schemes, in which hydraulic lines are eliminated and electronic signals (with redundant computers/real-time class-C MPX bus layout) modulate electrohydraulic or electric brake actuators at each wheel. Brake developers also dream of integrating ABS/TCS/VDC with future Intelligent Transportation Systems (ITS) concepts such as intelligent cruise control systems and automatic collision avoidance systems. These could allow safer, closer spacing of vehicles (e.g., "platoons" of vehicles, perhaps on automated highways) with much more efficient movement of passenger cars.

The 100-year evolution of passenger car brake technology appears to be rather revolutionary in its totality. The key word has been "safety" throughout; brakes must do their job every time, with no questions asked.

A History of the Passenger Car Tire: Part I

William J. Woehrle
Automotive Engineering Management Services, Inc.

It is virtually impossible to imagine what life would be like today without the automobile. In much the same sense, it is impossible to imagine what the automobile would be like without pneumatic tires. In its own unique and perhaps underestimated way, the pneumatic passenger tire plays as vital a role in shaping our transportation and, therefore, our civilization as does the internal combustion engine.

Philosophically, the tire is quite a paradox. It literally makes the passenger car what it is, while at the same time producing the greatest limitation to the vehicle's performance. Depending on one's perspective, the tire is the first or the last step in starting, stopping, and changing direction of the automobile as it travels. It produces cushioning and damping properties that cannot possibly be matched by other vehicle components. Yet, these tire performance characteristics combine to produce the limits for the automobile. One could go faster, stop in shorter distances, and enjoy a

(Courtesy of Michelin Tyre Company.)

smoother and quieter ride if those tires were better. Indeed, tires are better, and they are getting better every day. Not coincidentally, so too does the automobile.

The Tire Needed the Wheel— And Vice Versa

When was the wheel invented? We are not sure. It probably occurred around 3500 B.C. The idea of using logs to roll objects probably evolved into shorter and shorter segments. These segments eventually became disks with something resembling an axle connecting them.

As the evolutionary process continued, this hard, wooden wheel would last longer and operate more efficiently if its outer edges were covered with a material different from that of the rest of the wheel. Initially, this was a harder wood or metal. Later it was learned that ride comfort was improved and road shock was reduced by using a softer material such as leather. The wheel, therefore, had an *attire* to enhance its performance. Through the years (or centuries), the term was shortened to *tire*.

The Pneumatic Tire—Its Birth and Evolution

It is said that necessity is the mother of invention and that success has many fathers (but failure is an orphan). In this case, it seems as if the father appeared twice, approximately 40 years apart, but for two different reasons or applications. Furthermore, the invention was met with the usual skepticism—in this case, even from the inventors themselves. More-over, it was surrounded by some intrigue and confusion, which ultimately led to the opportunities and progress needed to keep pace with the fast-growing automotive industry.

Robert W. Thomson (1822–1873) is credited with being the first inventor of the pneumatic tire. The application, of course, was the horse-drawn carriage. Thomson, evidently, was looking for ways to reduce the pulling effort of the vehicle as well as to improve its cushioning and impact absorption qualities.

Thomson experimented with several constructions essentially involving a leather-covered outer casing with an internal, rubber-coated canvas air chamber. The leather covers were stitched or sewn together. He also ex-perimented with rivets, which he felt would improve traction. In this sense, it is interesting to note that the first tires had studs.

It is also interesting to note that the first tire test probably was, in fact, a rolling resistance test. Thomson claimed a 60% improvement on a smooth surface and as much as 300% improvement on rough roads.

Thomson's first patent, granted in England, was dated December 10, 1845. He also was granted patents in France in 1846 and in the United States in 1847. Unfortunately, it appears as if the horses were the most impressed with their reduced workload, but that they were not vocal enough to cause widespread sales of this new product. Accordingly, the pneumatic tire collected dust, and Thomson went on to pursue other ideas and ventures.

John Boyd Dunlop (1840–1921) emerged as the second father. The appli-cation was the cycle. Again, the test was that of rolling resistance. Dunlop, a veterinarian, demonstrated, in the courtyard of his veterinary establishment, a comparison of his pneumatic-tired cycle wheel along with a conventional solid-tired one. First he rolled the solid-tired wheel across the courtyard, and it wobbled to a stop before reaching the other side. Next he "tested" the pneumatic assembly with presumably the same amount of "push." It traveled the entire length of the courtyard, bounced against the gate at the far end, and began to roll back.

The first "vehicle test" occurred on the night of February 28, 1888, and involved his 10-year-old son, Johnny, and Johnny's tricycle. Dunlop applied for a patent on July 23, and it was granted on December 7 of the same year. This time it appeared as if the invention would catch on, and Dunlop and his colleagues set about to exploit the market with their new invention and the monopoly they envisioned. However, in 1890, when they discovered Thomson's previous patent, things seemed to unravel rather quickly. Instead of a monopoly and notwithstanding a short-lived cartel that was established in the United States, hundreds of tire companies emerged over the next few decades.

This patent disappointment or opportunity (depending on one's perspective) triggered a flurry of activities in both manufacturing and development. As with any invention, there was considerable room for improvement. Assuredly, the first pneumatic tire did not roll very far before the first flat tire emerged. Repair and service quickly became issues.

Early tire changes were difficult jobs. Here is one at a 1911 car race.
(Courtesy of Continental AG.)

Thomson's tire had as many as 70 bolts requiring removal for service. Dunlop's versions had so many drawbacks that he, himself, viewed the pneumatic tire being replaced, superseded by something else within a few years. These shortcomings spawned the improvements. The first wire-beaded tire was invented by Charles Kingston Welsh in 1890. The tire continued to be essentially tubular in shape, but the wires were placed in a position where one would expect them to be, and the corresponding contour was made in the rim to facilitate and secure the positions of these bead wires.

Later that same year, this concept was further refined by William Bartlett who was awarded the first patent for a detachable pneumatic tire. This ultimately became the standard construction for automobile tires as they are known today. His detachable tire ended at the bottom of each sidewall with a stiff portion including bead wires which engaged a curved-over lip or flange of the rim. Hence, the term "clincher" was used.

The notion of a detachable tire is what prompted the Michelin brothers to enter the tire business. Already involved in the leather and rubber fabricating business, they responded to the needs for quickly and efficiently repairing or replacing bicycle tires that had gone flat from punctures. Their patent was granted in 1891 and, with that, tires became their primary business. They pioneered with the first automobile tire in 1895. Edouard Michelin developed the tire, and his brother Andre demonstrated the product in a race from Paris to Bordeaux. Their production and marketing of the tire began in 1896.

In the United States, the first pneumatic automobile tire was produced by a company owned and run by a Civil War veteran and doctor named Benjamin Franklin Goodrich. Of course, Goodrich's application did not include pneumatic tires. His rubber-goods company was moved from New York to Akron, Ohio, to escape some of the cutthroat competition. An auto firm run by Alexander Winton of Cleveland ordered a set of pneumatic rubber tires. Reluctantly, B.F. Goodrich complied, but requested payment in advance.

The Automobile: A Century of Progress

Edouard and Andre Michelin were early pioneers, developing the first automobile tire in 1895. They demonstrated the product in a race from Paris to Bordeaux. (Courtesy of the Michelin Tyre Company.)

As early as 1891, the first pneumatic tire in the United States was produced for bicycles by a firm called the New York Belting and Packing Company. This was one of the firms bought by Charles R. Flint to form the famous rubber trust which later became known as the United States Rubber Company.

In the same way as it was a natural step for a bicycle tire manufacturer to become an automobile tire manufacturer, more than 200 bicycle companies chose to become automobile manufacturers. The famous example was Gormully and Jeffrey Manufacturing Company of Chicago, which later became American Motors Corporation. The original Rambler was a famous bicycle. Also, many of the cycle manufacturers produced their own tires. An exception, which was an early and successful specialist, was the

Morgan and Wright Company of Detroit, which at one time held an estimated 70% of the cycle tire market in the United States. This company was bought by the United States Rubber Company in 1914.

As automotive technology and manufacturing capacity raced beyond the horseless carriage era, the tire industry did its best to keep pace. The breakthroughs and improvements were on all fronts: construction, textiles, materials, and manufacturing.

Tire Construction Technology

The original reinforcement material for the pneumatic tire was square-woven fabric—first linen, then cotton. Cotton was the mainstay cord material through the Great Depression, reaching its high water mark with annual consumption of nearly a million bales.

However, the shortcomings of square-woven fabric were recognized quite early. The continued and inevitable distortion caused by a deformed pneumatic tire under load caused an incessant sawing action of the "weft" and "warp" cords in square-woven fabric. Understandably, as loads and torques increased, so did the problem.

The breakthrough was accomplished by John Fullerton Palmer who started the Palmer Cord Tyre Company in England and, in rapid sequence, the B.F. Goodrich Company in the United States with its "Silvertown" tire. The accomplishment involved constructing the fabric in a way such that the cross or weft cord appeared occasionally or not at all. In the manufacturing process, the cords were kept in place by a layer of rubber. This also facilitated application of the plies in a manner that produced even tension across the rounded contours of the tire. Another benefit was being able to twist and bundle the yarn in a fashion that was much stronger than what could be accomplished in woven cloth.

In the progression of textile materials, the next major step occurred in 1931 with the first manmade material—rayon. Made from wood pulp, the material had to overcome significant adhesion difficulties—as was the case with virtually every superseding cord material. The marketplace saw the first rayon tire in 1938. By 1955, it had virtually eliminated cotton as a tire reinforcement material.

The next step occurred with nylon—the first completely synthetic fiber. It made its debut in the marketplace in 1947. What ensued was a classic battle of nylon versus rayon, which lasted for several years. Nylon was considered to be stronger, less resistant to heat, and produced less heat; however, it did not offer the dimensional stability of rayon, and among the most notable manufacturing changes triggered was the post-cure inflation to achieve better uniformity as the tire cooled.

The nagging annoyance which ultimately doomed nylon as a passenger tire reinforcement material was "flatspotting." This deflection set from a stationary car and tire would remain for a brief travel distance, producing an annoying thump. It was especially annoying as an original equipment tire having the potential to all but ruin the "first-impression test drive." After the tire warmed up and regained its original round shape, the thump would disappear. However, the fear was that the negative impression would have already been created and the marketing damage would have been done.

The answer occurred with polyester in 1962. It had the strength of nylon, but none of the flatspotting properties. It quickly caught on and remains the body-ply cord of choice for passenger car tires today.

As for belt material technology and its associated progress, one must first understand the belt's significance. Dating back to the first cord tires, belts may have been handy, but they were not necessary. The reason for this is that the reinforcing materials were arranged in diagonal or bias-angle plies. Such cords would run diagonally from bead to bead. In the manufacturing process, if such plies were assembled on a conventional cylindrical drum, the bias-angle plies would "pantograph" when the tire was shaped. The most significant breakthrough in tire technology occurred with radial-ply construction. In a manner strangely similar to the pneumatic tire itself, it seemed to occur well ahead of its time. The first patent was granted in 1913, and the inventors were Christian Hamilton Gray and Thomas Sloper. However, it remained dormant until Michelin introduced the Michelin X radial-ply tire in 1948. Again, a rolling resistance test was key to its message, together with claims of longer tread life and improved handling.

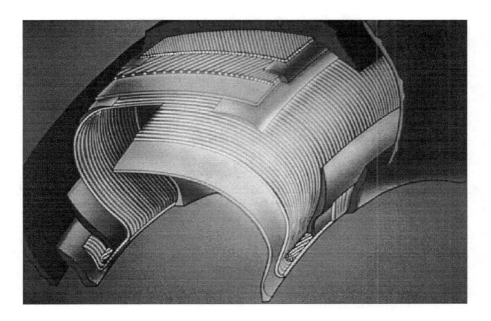

A significant breakthrough in tire technology occurred with the introduction of radial-ply construction. Shown is the Michelin X radial-ply tire of 1948. (Courtesy of the Michelin Tyre Company.)

A radial-ply tire gets its name because the body-ply cords run directly across from bead to bead. Therefore, in the sidewall region, they appear as radii of the circle. Circumferential strength is provided by a belt consisting of cords running at a very small angle to the circumferential direction of the tread and only under the tread (not down the sidewall).

The original belt material patented by Michelin in 1946 was steel wire. Again, this posed a major challenge for adhesion to rubber. Nevertheless, as this challenge was successfully met and the construction perfected, the performance benefits were nothing short of overwhelming.

The Michelin patent, together with adhesion and manufacturing challenges, motivated other tire manufacturers to consider belt materials other than steel. Hence, rayon belted radials emerged in the marketplace, followed by fiberglass belts. A recent newcomer in belt material is aramid. This synthetic fiber, also known as Kevlar and marketed as "Fiber B" by

DuPont, is claimed to be up to five times stronger than steel—pound for pound. It is most likely to be found in high-performance or high-speed radial-ply passenger tires.

The radial-ply tire indeed revolutionized the tire industry. Tire factories closed, and companies went out of business in its wake. The winners found themselves producing tires that were far superior to the bias-ply predecessors.

A logical interim step in this "revolution" was the bias-belted tire. Pioneered by Armstrong in 1965, it consisted of a bias tire, but with additional reinforcing belts made of fiberglass. The other tire companies quickly joined in with this popular construction consisting of polyester body plies and fiberglass belts. Later bias-belted tires emerged with steel belts. The performance advantages lay somewhere between the bias and radial construction. The primary advantage was that much, if not most, of the bias tire manufacturing equipment could remain in use and the resultant tire prices remained attractive.

However, radial-ply construction was not to be denied, and through the 1970s the bias-belted tire gradually all but disappeared from the marketplace.

Rubber and Compounding Technology

The key technological breakthrough which began this whole program called the pneumatic tire was, of course, the vulcanization of rubber. Charles Goodyear discovered this process, quite by accident, as he admitted in 1839. Several years were spent improving the process, and he was awarded a U.S. patent in June 1844. The vulcanization process essentially involved the addition of sulfur and heat which rendered the physical properties of rubber far more stable. Prior to vulcanization, rubber would become too soft and sticky in warm ambient temperatures and too brittle in cold ones. Instead of tires, Goodyear was focusing on footwear and other rubber products. However, it was inevitable that this vulcanized rubber was the material of focus from which a product called a pneumatic tire could flourish.

Rubber, of course, is a naturally occurring product, and its cultivation was started in 1876. In fact, the world relied on natural rubber—with virtually no suitable substitute—for nearly a century. Even today, natural rubber enjoys a robust demand.

Carbon black was added to rubber to produce desired improvements in strength and hardness, beginning in 1904. Sidney C. Mote of the India Rubber, Gutta Percha and Telegraph Works in Silvertown, England, is credited with this accomplishment. Again, this was probably an idea before its time, and it was not until 1912 in the United States that the Diamond Rubber Company and B.F. Goodrich used this addition to rubber. The use of carbon black did not become widespread until the early 1920s. Prior to that time, tires were white instead of black.

Another significant improvement occurred in 1906 when George Oenslager of the Diamond Rubber Company (B.F. Goodrich) developed accelerators to shorten the curing time of rubber. This produced a major improvement in product quality and significantly reduced cost.

As tires became longer lasting and longer wearing, the oxidation problem of rubber became

The first all-synthetic rubber tire produced in the United States was by B.F. Goodrich in 1940. This advertisement appeared in the Saturday Evening Post *(October 5, 1940) and* Life *(October 21, 1940). (Courtesy of University of Akron archives.)*

more noticeable. This led to the development of organic age resistors or anti-oxidants in 1924. Harold Gray of B.F. Goodrich and Sidney M. Cadwell of the United States Rubber Company were the primary technicians.

The first major step toward synthetic rubber was taken in Germany in 1931. Eduard Tschunker and Walter Bach successfully produced "Buna S" by using an emulsion polymerization of butadiene and styrene. The butadiene monomer "BU" is the synthetic near twin of natural rubber isoprene. ("Na" is the symbol for the element sodium, which served as the catalyst.) In 1933 Semperit of Austria produced the first synthetic rubber tire. The first all-synthetic rubber tire produced in the United States was accomplished by B.F. Goodrich in 1940.

As the winds of World War II approached, it became abundantly clear that, given the geographic location of natural rubber plantations, an all-out effort to enhance and broaden the development and manufacture of synthetic rubber would be a key to America's success. Accordingly, the U.S. government established national committees, bringing together industry, government, and academia, and essentially waiving all competitive restriction to accelerate the development process. The result was a recipe for GR-S, developed in 1941. This was the forerunner of SBR (styrene butadiene rubber).

The manufacturing capacity of this synthetic rubber was brought onstream as quickly as possible during the war years to keep pace with the development accomplishments. Before the end of the war, its supply was no longer a problem.

A further enhancement of SBR occurred in 1948 when low-temperature polymerization was perfected by a task force headed by Carl S. Marvel of the University of Illinois. This triggered additional expanded usage such that SBR today remains as the standard of synthetic rubber for tire manufacturing.

Another advancement occurred with the development of polyisoprene synthetic rubber which was introduced in 1954 by B.F. Goodrich and Firestone—simultaneously and independently. It was considered to be the synthetic rubber that most closely resembled the properties of natural rubber.

Enhancements continue even today as challenges for improved performance are unending. Improved traction, treadwear, and rolling resistance continue to be held as collective goals which, together with competitive pressures, keep the compounders busy searching for ways of improving one property while not sacrificing others.

A History of the Passenger Car Tire: Part II

William J. Woehrle
Automotive Engineering Management Services, Inc.

It is virtually impossible to imagine what life would be like today without the automobile. In much the same sense, it is impossible to imagine what the automobile would be like without pneumatic tires. In its own unique and perhaps underestimated way, the pneumatic passenger tire plays as vital a role in shaping our transportation and, therefore, our civilization as does the internal combustion engine.

Philosophically, the tire is quite a paradox. It literally makes the passenger car what it is, while at the same time producing the greatest limitation to the vehicle's performance. Depending on one's perspective, the tire is the first or the last step in starting, stopping, and changing direction of the automobile as it travels. It produces cushioning and damping properties that cannot possibly be matched by other vehicle components. Yet these tire performance characteristics combine to produce the limits for the auto-mobile. One could go faster, stop in shorter distances, and enjoy a smoother and quieter ride if those tires were better. Indeed, tires are better, and they are getting better every day. Not coincidentally, so too does the automobile.

Manufacturing and Process Developments

Because a pneumatic tire is a rubber membrane with reinforcements of textile cords, the first task was to combine these two basic components efficiently in the manufacturing process. This was first accomplished in 1836 by Edwin M. Chaffee of Roxberry, Massachusetts. He developed a machine capable of applying rubber directly to fabric without solvents, producing a sheet of rubber of uniform thickness. The machine he used was called a calender. It remains in use today and is the mainstay of any tire factory.

After all of the tire components are assembled and the tire fully built up, it is ready for vulcanization or cure. Initially, this was done in open steam autoclaves. The tire would be held in place with some ring-shaped molds which, in turn, were held in position by a wrapping of canvas bandage. In 1896 H.J. Doughty invented the curing press which included a mechanical means of supporting the tire from the inside, and by applying pressure or inflating the tire from the inside, forcing it against a mold pattern to produce elaborate tread designs and sidewall treatment (e.g., lettering). The vulcanization or curing proceeded while such molding was occurring. Again, this basic process, along with countless improvements, is in use today.

In 1916 F.H. Banbury developed a machine to speed the process of mixing chemicals and rubber in its uncured state prior to the calender operation or the making of other tire components. This was considered a major milestone in processing, and the Banbury mixer remains a mainstay in tire factories worldwide today.

The radial-ply tire posed as much or more of a challenge to the factory as it did to the engineer or chemist. For a conventional bias-ply tire, all of the components are assembled in a one-stage manufacturing process prior to curing. That is, the tire is assembled on a single machine (usually a building drum). As the radial-ply tire gained a foothold and flourished in the marketplace, countless improvements in the manufacturing process occurred to keep pace. Moreover, the designs and key features of the machines were unique to the innovating tire manufacturer and were closely guarded secrets. With the fast-paced changes and improvements via electronics, computers, and robotics, it is safe to assume that modern tire factories are keeping abreast of these opportunities and making changes accordingly.

Shape, Design, and Styling Trends

Just as the automobile has changed in appearance through the years, so has the passenger tire. Yes—today the tire still is a rubber-coated, fabric-reinforced, pneumatic pressure vessel. Yes—the vehicle it is supporting still has four wheels and is powered, in most cases, by an internal combustion engine. Beyond those similarities, the changes are dramatic.

The change in the tire size and profile has been a steady progression to lower and wider. A tire's cross section or profile is defined by its aspect ratio—a height-to-width ratio. The height is the distance from rim to the tread; the width is sidewall to sidewall. This ratio, expressed as a percentage, has continued to march downward through the years.

The earliest tires had an aspect ratio of 100%. This continued until the early 1920s. In 1923, Michelin introduced the balloon

In the early 1930s, General Tire Company promoted balloon tires effectively. (Courtesy of Continental General Tire, Inc.)

An early balloon tire from Continental AG. (Courtesy of Continental AG.)

tire in Europe, and Firestone was the pioneer in the United States. The idea did not catch on heavily until the early 1930s when General Tire Company promoted it effectively. Having an aspect ratio of 98%, the tire also featured nearly twice the tread width and, under normal conditions, only 28 psi inflation pressure was required. This was approximately half that of the tire it replaced; thus the key benefit was a much softer ride.

This led to the super balloon tire which emerged after World War II and had an aspect ratio in the low 90s. The downward trend continued below 90% with the adoption of low-profile tires in the late 1950s. Soon a super-low-profile tire with an aspect ratio of close to 80% appeared in the marketplace.

Passenger tire sizes were expressed in terms of a section width in inches followed by the rim diameter in inches. With this movement toward lower aspect ratios, the new sizes had larger numbers but virtually no change in load-carrying capacity, e.g., 7.75 x 14 replaced 7.50 x 14. Foreseen as a continuation of these trends together with an attempt to avoid confusion, the alphanumeric system of passenger tire sizes was adopted. This was referred to as a "load-based" system as opposed to its predecessor, a "dimensionally based" system of passenger-car tire sizes. It was to convey the important message that, as aspect ratios became lower and tires became wider, the load-carrying capacities would be the same. The load increments were established, and a letter of the alphabet was assigned as the first character in the size nomenclature. The letters A through L (except for I) were used, representing low to high loads for passenger cars, respectively. The next two characters in the nomenclature were numbers for the aspect ratio, expressed as a percentage.

In the late 1960s, 78 Series and 70 Series sizes were adopted and soon approximately 60 Series sizes appeared. This trend toward lower aspect ratios, keeping pace with a flurry of activity in radial-ply development, served as the answer to steadily increasing demands for higher speed and higher performance in passenger tires.

The trend toward lower aspect ratios continued into the 1970s. However, shortcomings in the alphanumeric system of sizes had to be addressed. First, with the introduction of the Chevrolet Chevette, a size smaller than A was needed. Second, interest in converting to a metric system was growing in the United States. Last but not least, there was a desire to establish a common system of tire sizes and to establish a worldwide standard. Thus, the P-metric system of sizes was born. Reflecting a return to a dimensional-based system of tire sizes, it also included aspect ratio, and a provision for meaningful increments was incorporated.

Aspect ratios continue to drop to as low as 35 Series tires, which have been recently introduced. In addition to high speed and handling performance, another advantage of the low-aspect-ratio sizes is the opportunity to use larger rim diameters and therefore larger brakes.

Tread design trends, of course, have focused on improvements in traction on surfaces other than dry pavement. For wet-traction performance, the simple yet never-ending challenge to the tread design engineer is to provide sufficient and efficient paths for water to escape and "get the rubber on the road." Circumferential grooves with more "see-through," the use of shoulder slots, and, lately, wide center grooves have all represented progressive accomplishments for this performance characteristic.

As tires and treads became wider, there was an opportunity to increase the number of ribs from five to seven, and, in some cases, as many as nine. Again, wet-traction performance and resistance to hydroplaning were seen as benefits. Similarly, the highway officials determined that the cutting of longitudinal grooves in the pavement, spaced 0.75 inch apart, would also reduce the risk of hydroplaning by improving water drainage. Therein occurred the clash! When a sufficient number of these highway grooves interfered with the tire tread grooves, a small but annoying lateral displacement of the vehicle occurred. This has been called "squiggle," "groove wander," and probably some other terms unfit to print. This hastened a retreat of sorts, back to five-rib designs which remain the norm for popular passenger tire sizes today.

For snow-traction performance, the key is to optimize or maximize the lateral projections of grooves and sipes, especially in the outer ribs of the tread design. For tread designs that would do well in this area, the major challenge or tradeoff is typically in the area of uniform and lasting

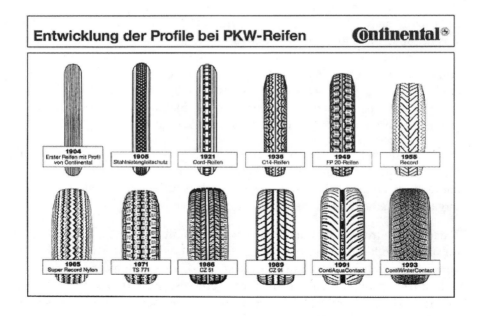

Continental tires throughout the years. (Courtesy of Continental AG.)

treadwear as well as noise. The first snow tire was offered by Continental of Germany in 1909. Through the 1970s, the conventional practice was to change from highway tires to snow tires in the winter and vice versa in the summer. Enter the all-season tire, pioneered by Goodyear, which obviated the need for this twice-a-year changeover. Soon the rest of the industry offered their versions, and by the early 1980s the all-season tire had become well established on passenger cars.

Some vintage General Tire designs. (Courtesy of Continental General Tire, Inc.)

This refinement continued through the low-aspect-ratio and high-performance tires as the competitive pressures, together with widespread consumer acceptance, encouraged this progress. Today, "M&S," indicating such all-season applications, are the rule rather than the exception on passenger tire sidewalls.

The sidewall treatment has been a story of styling. Just as the car has gone through the eras of tailfins and chrome, the sidewalls have gone from wide white to narrow white, to multiple stripes, to white letters, to outlined white letters, and so on. Molded sidewall treatment (in white or black) included variations such as "spangles," diamonds, facets, and rings, as well as an endless variety of lettering. An example was the "General Streamline Jumbo" in 1934. For a brief period, colored sidewalls appeared in the early 1950s, with tires from the U.S. Rubber Company. Later examples included the U.S. Royal Tiger Paw and the Goodyear Blue Streak. Lately, blackwall has been the most fashionable. What goes around, comes around.

Yes, But Will It Hold Air?

How soon after the first pneumatic tire started rolling did the first flat tire occur? Probably no one knows for sure. It seems certain that Thomson and Dunlop could provide us with many interesting stories. The objective of holding air remains with us today. It is not hard to imagine that Thomson's rivets were problematic in creating punctures.

There was also the problem of how to inflate or deflate the tire. Dunlop's invention involved the use of a crude valve similar to that used in footballs. It was one way, i.e., the tire could be inflated but not deflated. The first major improvement in a tire valve occurred in 1891 when Charles H. Woods offered a two-way valve, allowing deflation as well as inflation. The next major step was taken by George H. Schrader in 1898. His product provided for a screw-in replaceable valve core. This quickly became the standard and remains in use today.

General Tire's Gen Seal sealant tire.
(Courtesy of Continental General Tire, Inc.)

The air retention properties of rubber were significantly improved with the invention of butyl synthetic rubber in 1937. Robert M. Thomas and William J. Sparks of Standard Oil of New Jersey (now Exxon) are credited with this accomplishment. Further improvements were made by Francis Key Baldwin with chlorinated butyl, hence the name chlorobutyl which is the mainstay for tubes and tubeless liners today.

The tubeless tire was invented in 1947 by Frank Herzegh, an engineer with B.F. Goodrich. With the tube-type tire, there was a risk of the tube squeezing through a hole, cut, opening, or worn fabric. In such incidents, the possibility of a rapid air loss (blowout) existed. The primary claim for the tubeless tire was that this risk was eliminated or at least minimized.

The use of steel belts—in bias-belted tires as well as radial-ply tires—offered a major improvement in road hazard protection. With road hazards such as broken glass, air loss problems became all but a distant memory. Yet, the nagging question remained—"What's to be done about that incessant nail?" A possible answer was put forth by Uniroyal and General with their sealant tires.

In somewhat of a different direction, answers have been provided which involve continued mobility, even after air loss. The "Goodyear Double Eagle Airwheel and Lifeguard" was promoted in 1938 as a reserve tire inside the tube. Both were inflated through the same valve. If the outer tire and tube failed, the inner tire held enough air to support the car until it could be brought to a "smooth, safe stop." An extension of such a "run flat" tire was offered by Dunlop in 1972. Numerous other examples have been evaluated, and the development activity continues.

There is a never-ending battle being waged against the road hazards—and producing tires that continue to do better against this onslaught. The ultimate victory will be viewed with the elimination of the spare tire!

Spare Tires—A Love/Hate Relationship

So the tire goes flat—then what happens? If you do not have the skills and tools to repair the tire on the spot, the next best thing is to change the tire and continue your travel.

In the early 1900s, during the period of the clincher tire, no one had yet championed the idea of having a mounted spare tire/wheel assembly stowed somewhere aboard. As flat tires inevitably occurred, the motorist was forced to deploy an elaborate set of wrenches, crowbars, levers, clamps, jack screws, and even a pickax was recommended to wrestle the tire free from the rim. Usually the same tire was repaired and put back into service on the spot if possible. The breakthrough of equipping the car with a spare tire/wheel assembly must have been well received by any motorist who had the former tire-changing experience.

Through the years, the spare tire continued to get in the way of stylists, packaging experts, and motorists who wanted to carry "just a little bit more baggage." Yet, as common as flats were, well into the 1930s the long-distance touring cars often were equipped with two spare tires.

The first major step toward relinquishing some trunk space occurred with the collapsible spare, marketed by B.F. Goodrich and Goodyear in the late 1960s. This was a bias tire which, when stored in the trunk, was uninflated and collapsed to an overall diameter slightly larger than the rim. An infla-tion canister was provided, and when placed into service and inflated, the tire expanded to roughly the same dimensions as the conventional tires. Although this invention required less trunk space, it offered no weight savings and, considering the addition of the canister, probably no cost savings.

The next major breakthrough came with the mini or compact spare, which was pioneered by Firestone in the late 1970s. This also was a bias tire and offered the triple benefit of space, weight, and cost savings. It has wide-spread usage today and is found in the trunks of most passenger cars. Also, versions are available with radial-ply construction.

The mini spare frequently did not get the respect it deserved. It was viewed as cheap, a crutch, and something that could not possibly work as well as the disabled product it was temporarily replacing. Yet it was and is a well-engineered product, and, on the occasions that it was placed into service, the motorist hardly could notice the difference in vehicle performance.

Moreover, the mini spare became somewhat of an effective spokesperson for the performance of today's passenger tires. Many automobiles run through their useful lives and are scrapped without the mini spare ever being removed from the trunk.

Rolling on and into the Future

Yes, the automobile has come a long way in the last 100 years, and the pneumatic tire has been with it for nearly every mile. As much as anything, mileage is the dramatic yardstick for measuring the improvement through the years.

In the beginning, the life expectancy of automobile tires was a few hundred miles. Today 40,000 to 60,000 miles are typically traveled before the tires are worn out.

Yes, the spare tire remains in the trunk. But it is brought out and put to use more for that pesky nail than for more catastrophic failures.

Today's tire runs quietly and rolls smoothly; now, it would help if the pavement were a little smoother. Experts tell us that the direction of tire technology is indeed more of the same. Longer wear and a smoother ride can be expected. Traction and handling will continue to improve; and yes, the elimination of flats and, therefore, the spare tire remains the significant target. It is not a question of if, but when, the spare tire will disappear.

As in the beginning, rolling resistance remains important today and will continue to be important in the future. Fuel economy demands and opportunities will continue to push the tire's rolling resistance to levels never before thought possible. As has always been the case, materials breakthroughs will remain the key to these improvements. From reinforcing cords, to adhesives, to new synthetic rubber compounds—advancements in all areas are envisioned.

Will the pneumatic tire survive? One might similarly ask—"Will the internal combustion engine remain, or will the automobile itself remain?" Frankly, no one has pondered anything reasonable as an alternative.

A Century of Automobile Body Evolution

Karl E. Ludvigsen
Photos courtesy of The Ludvigsen Library Ltd.

So significant is the auto body as a consumer of materials that it has attracted the attention of many materials suppliers, with the result that an unparalleled range of possible materials is available to the body designer of the 1990s. That the standard body material, steel, is also the most recyclable will pose new challenges to the body engineers of the next century who champion alternative materials.

Not to be overlooked is the role of the auto body as an expression of the personality of the company that produced it. Both stylists and aerodynamicists have made strong contributions to the evolution of these personalities. The mission of the auto body as a structure that must protect its occupants is being understood and mastered more and more comprehensively at the end of this century.

Daimler of Stuttgart was a pioneer of several engineering techniques that helped to establish the concept of the automobile body as we knew it for at least the first 50 years of the industry. The 1901 Mercedes introduced by

The 1901 Mercedes set the standard for future chassis and body design.

Daimler had a vertical honeycomb-core radiator at the front of the car, a completely new concept. Before the Mercedes, automobile engines were cooled—if at all—by arrays of finned or gilled tubes carried at the front of the car or below it. These ugly tubes were a necessity rather than an attractive feature of the automobile.

With the invention of the honeycomb radiator by Wilhelm Maybach of Daimler, the way became clear to fit radiators of different shapes and sizes to give distinctive identities to various types of automobiles. Not surprisingly, the first action of many car-makers was to copy the Daimler Mercedes radiator design.

The exposed radiator core surrounded by a distinctive shell continued to be used through the 1920s. The next step was to place a grille or screen in front of the core that protected it from stone damage and could also be used to control cooling airflow and thus engine temperature. Car-makers were quick to develop distinctive designs for these grilles, which became more integrated with the hoods and cowls.

In the 1930s, the further enclosure of the radiator into the bodywork progressed rapidly. Cars identified with this trend were Adler, Skoda, Steyr, Peugeot, Tatra, and Chrysler with its Airflow. In February 1937, *The Autocar* predicted: "It will not be long before the radiator will be a

The woodworking shop was a major contributor to car body construction well into the 1920s.

completely forgotten unit, quite lost behind its prison of horizontal or vertical slats. The day when each car had its own jealously guarded radiator shape has irretrievably gone."

Body design was also affected by another important innovation in the 1901 Mercedes by Daimler: its use of a pressed-steel chassis frame. Nineteenth-century autos had short wheelbases precisely because designers had not yet discovered how to contain springing and drive stresses within the early chassis frames. Many frames were made of wood with steel or iron reinforcements.

The introduction of the pressed-steel frame, which could be shaped to accommodate the varying stress levels at different points along its length, allowed automobiles to become longer. Occupants could be seated lower in a long chassis. Side door openings could be provided for the rear passengers, replacing the old rear-tonneau entrance.

With the creation of the steel frame, the way was open to extreme and impressive body length for the most expensive cars. It was obvious that a long car cost more money to buy and also cost more to store and maintain. Thus sheer length became a hallmark of the early prestige cars. Wheelbase lengths of leading cars of the classic era are shown in Table 7-1.

Table 7-1
Prestige-Car Wheelbase Lengths

Bugatti Royale	4300 mm
Daimler V-12	4140
Bentley 8-Liter	3962
Cadillac V-16	3912
Duesenberg J/SJ	3899
Mercedes-Benz Grosser	3886
Rolls-Royce Phantom II	3810
Horch 12	3750
Maybach SW38	3680

Body engineers were given a substantial challenge by these magisterial wheelbase lengths. Body-engineering techniques for cars were literally carried forward from the days of the horse-drawn carriage. People were accustomed to instruct builders of carriages to design and create special bodies for them to meet their requirements. These coachbuilding skills, and also the personal customer contacts of the relevant companies, were carried over directly to the new horseless-carriage era.

In Britain, for example, Hooper & Company was founded in 1805. For more than 130 years, it held the warrant to supply coachwork to the British Royal Family. Britain's Barker was founded in 1710. Its first car body was constructed for Charles Rolls in 1905. The Mulliner line stems from the coachbuilding business of Arthur Mulliner, founded in Northampton in the 18th century. In 1900, H.J. Mulliner broke away from the family business to found a car body-building company in London.

American coachbuilders could command comparable lineages. Brewster & Company, which traced its origins to 1810, was acquired by Rolls-Royce in 1926 when the British auto-maker began production in the United States. Judkins of Massachusetts was founded in 1857, and Derham of Pennsylvania was another 19th-century firm.

The body of this Riley used fabric over a wood frame.

LeBaron was among the other coachbuilders catering to wealthy Americans (designer Hugo Pfau: "Certainly, with one or two exceptions, none of our customers at LeBaron had less than several million dollars."). The large firms competing for their trade were Locke, Rollston, Fleetwood (bought by Fisher Body in 1925), Brunn, Willoughby, Holbrook, Murray, Dietrich, Murphy, and the French Weymann company, which set up a plant in Indianapolis to produce bodies using its patented system of a light, flexible fabric outer covering supported by a wood frame.

In Italy, many of the early coachbuilders traced their origins to the horse-drawn era. One that remains active is Bertone. Another was Sala, which was not destined to survive to the most creative period. Yet another was Castagna, an authentic inheritor of the original coachbuilt traditions that contributed some of the most striking bodies of Italy's cars. While Zagato built the most sporting Alfa Romeos, Castagna produced the most elegant Alfas. Castagna was, in fact, revived in 1995.

In the heyday of the coachbuilt automobile, the car owner's mechanic or chauffeur often played a key role in deciding what chassis to use and which coachbuilder to clothe it. After gathering his opinions, the owner would

Open touring cars could be well enclosed for bad weather with the help of side curtains all around.

visit the prospective coachbuilder, or the latter might visit him. Sketches for proposed designs would be prepared to meet the owner's special requirements. Cost and timing would be established, and work would be put in hand.

This led to the customer having a strong influence on the way in which car designs evolved. Many such cars, styled to meet particular customer requests, set new patterns in auto design. The counterpart of such customer interaction exists today in the form of many specialized companies that produce restyled versions of cars such as Jaguar, Mercedes-Benz, and BMW to meet their owners' special requirements and offer additional distinction. If he/she was satisfied with the chassis, it was not unusual for an owner to have it rebodied from time to time. Equally, favored and treasured bodies were sometimes mounted on new chassis as car technology evolved.

As with this Hispano-Suiza, luxury cars were fitted with temporary bodies for final testing before their proper bodies were fitted.

There were sharp differences in character and approach in the methods used by body-building companies in various countries. As early as 1912, for example, an American visiting Europe noted that "the French designers are all for graceful and consistent design. In England the whole idea seems to be practicability and utility."

The French achieved their design flair by drawing the bodies full-size on large sheets of paper, an adaptation of an Italian concept. Among the French houses, Franay in particular was regarded as producing outstanding designs, deriving from its earlier tradition as a builder of horse-drawn coaches.

The British coachbuilders were viewed as likely to adapt their auto bodies more to the requirements of the chassis than to any aesthetic sensibilities. This meant that British bodies were viewed as practical and serviceable, but less attractive than those built in France.

An attribute of the car established in its earliest years was that it should glisten and gleam, in both its paintwork and its polished metal parts. Originally nickel-plated, the latter acquired even more gloss when they became chrome-plated. Traditionally more conservative models, such as town cars, were painted black or another dark color. This carried the clear implication that the owner was wealthy enough to afford the regular services of a chauffeur or other menial who would wash and clean the car. This was especially prestigious in the years before roads were fully paved.

Custom body builders were asked to paint cars in colors particular to the purchaser. Hibbard et Darrin of Paris, for example, was asked to paint a Rolls-Royce body the shade of a stocking furnished by a movie star. The lady was not happy with the first attempt, and the car had to be repainted to meet her specification.

Before the introduction of cellulose finishes, it could take as long as six months to paint the body of a luxury car. Even for an ordinary car, a man-year of labor was required to produce a fully finished body. Around 1910 to 1912, a skilled crew could require approximately eight weeks of working days to make a first-class application of the required finish materials. Painting a quality car at that time required successive coats of lead-color, filler, stopper, stain, ground color, and body color, covered and protected by flatting and finishing varnishes. Drying times were needed between these coats, and time also was needed for the intermediate operations of rubbing down, flatting, hardening-off, and lining-out.

The finest coachbuilders were experienced in applying the traditional paint and varnish system to produce a deep and attractive finish. The surfaces were not durable, however. Cars had to be varnished annually or repainted frequently. Thus the new cellulose paints were welcomed for their lack of maintenance even though they lost the deep, rich surface appearance of the paint and varnish systems.

Cellulose-based paints for cars were introduced during the 1920s as a result of research by DuPont in cooperation with General Motors. They reduced preparation times for bodies from weeks to hours, and eventually to minutes. The first car to be sold with Duco paint was a 1924 Oakland model.

The first applications in volume production of what we now call "metallic" paint occurred in the mid-1930s. Graham is credited by historian Jeff Godshall with its introduction as early as 1932. Fish scales were used at first to give the effect; later the addition of 5–10% of finely ground aluminum to the paint gave the desired iridescent appearance. GM began using the paint widely in 1937.

In the late 1950s, acrylic resins were introduced for automotive paints, substantially improving durability. Modern developments have included water-based paints to reduce emissions associated with auto painting.

An important improvement in body finishing was the introduction of electrodeposition methods to ensure complete application of primer. First used in the 1960s, electrodeposition was soon widely applied. By 1983, it was used on 90% of all cars.

Improved priming methods, as well as better sound-damping systems, were needed to cope with two of the problems caused by the new-fangled all-steel auto bodies: increased corrosion and resonance. These efforts were well worth it, for the all-steel body brought a revolution in productivity to the auto industry. André Citroën credited it with helping him accelerate his production from 30 to 50 cars per day to 400 to 500 in the 1920s.

The art of the all-steel auto body was immeasurably advanced by Edward Budd as producer and manager and Josef Ledwinka as engineer, starting when both were working for Hale & Kilburn of Philadelphia, makers of rail carriages. After making some all-steel bodies for Hupmobile in 1912, Budd set up his own business in that year.

On June 17, 1914, Ledwinka was granted U.S. patent 1,143,635 on an all-steel, all-welded auto body. Around this primary patent, Budd wove a network of improvements and advances. The Dodge brothers, setting up their own auto company, contracted with Budd to supply their bodies. In 1914, Budd began to supply steel-framed bodies to Dodge, replacing the traditional oak and ash frames. Output accelerated in 1916 with the introduction of an all-steel open tourer body. The technology was still in its infancy; each body required approximately 1200 separate stampings. Steel bodies moved into the high-priced car category in 1926 with their adoption by Jordan.

By 1919, Dodge was also offering closed four-door steel bodies. Hudson created a sensation in 1921 by introducing a Budd-built enclosed Essex sedan at only 25% more than an open tourer; the closed-car era was beginning. By 1927, the production share of closed cars in America, which had been 17% of output in 1920, had soared to 85%.

Budd's concepts and equipment were adopted with enthusiasm in France by André Citroën, most significantly in his 1925 10CV type B10 sedan. In Germany in 1926, Ambi-Budd was formed as a Berlin-based joint venture to produce and license all-steel bodies; BMW was an early customer. With Morris (who later withdrew) in Britain, Budd set up the Pressed Steel Company to exploit the Budd concepts.

The term "all-steel body" was a misnomer in the 1920s because rooftop panels were still a fabric, such as Textileather, over a frame of wood and chicken wire. Designers were concerned, and rightly, that a steel enclosure

1925 Citroën—its all-steel body was designed and built in cooperation with Budd.

would increase interior noise. However, by the early 1930s, the steel roof was a reality as well. GM hailed it as the Turret-Top when it was introduced on most of its 1935 models.

Thus, through the 1930s, the use of fabric-covered roof surfaces became an identifying mark of a custom-built car. To set their cars apart from the volume-produced models, coachbuilders would use a fabric roof. Similarly, when makers of volume cars wanted to offer models that aspired to higher levels of prestige, they would cover their steel roofs with a padded fabric top.

The use of a fabric-covered roof to imply enhanced prestige continued well after World War II, especially in the United States and Britain. However, car stylists object to such fabric roof coverings and have sought to design their cars in such a way that it is impossible to have them retrofitted. This has not deterred many U.S. dealers, who still arrange to have them installed, however inappropriate their designs, by aftermarket companies.

In parallel with the development of all-steel bodies ran the creation of press tools and systems for making their panels. An important development in the late 1920s was the invention of the first sheet-steel press tool by Artz in Germany. The Artz system used hydraulics to press a die against a sheet of steel that was held at its ends.

This early development was followed in the 1930s by the invention of the matched-die press of the type that remains in use today. This at last opened the door to the rapid production of steel body parts of identical design. It also tended to increase the cost of tooling and thereby to introduce new scale-economy requirements for entry into volume automobile production.

The ability to form more complex panels allowed the number of press parts per car to be reduced. Bodies in Europe that initially required 500 parts could be built with only 120 by 1930. The 1930s also saw the introduction of modeling in clay to help car-body designers and the new profession of stylists to master the more rounded forms made possible by matched-metal-die stamping.

The Automobile: A Century of Progress

Among the leaders in the use of full-scale modeling in clay was Harley Earl, who was asked by GM's Alfred P. Sloan, Jr. to establish the company's Art and Color Section in 1927. Strongly backed by Sloan, Earl was able to demonstrate the value of an approach to car design that set fine and distinctive appearance as a specific objective. Other pioneers of this new concept were the artists who began working at Chrysler in 1925 and industrial designers Raymond Loewy (Hupmobile) and Walter Dorwin Teague (Marmon).

One objective of the stylists became the smoothing and "perfecting" of the shape of the automobile. They wished to eliminate the mechanistic forms and odd junctions of surfaces and body sections that had been considered satisfactory by the engineers. By taking such steps, however, the stylists were also removing some of the signs of character, personality, and identity that had survived to distinguish important car brands since their origins.

The era of the stylists also brought a more frantic pace of change to designs from year to year. This rapid pace of change, particularly in the United States in the 1930s, made it increasingly difficult to establish and maintain a particular appearance and character. At the same time, however, bold new brand characters, using modern design idioms, were established by the early 1940s for such makes as Cadillac, Buick, and Chrysler.

Vital to the success of the process of deep-pressing steel parts and making frequent design changes was the ability to create the needed dies. In 1921, wrote historian Stan Grayson, Edward Budd "chanced to meet a Brooklyn tool manufacturer on a train. The man, Joseph Keller, had invented and built a mechanical engraving machine suitable for use in making automotive body dies. Budd bought the first machines, was able to lower tooling costs, and saw the machines become standard in the industry." Thus was born the Keller method of copying a die model to produce die sets.

Methods of assembling steel body panels were also improving. The introduction of electric spot welding was a key innovation. Pioneers Dodge and Budd worked on this jointly and introduced the first spot-welded body in 1927 for the Dodge Victory 6. Spot-welding has since become the standard method for steel body assembly.

Lancia was the first to unify the frame and body of a car in a practical manner with the 1922 Lambda.

In the 1990s, Volvo began testing a method of pressure-seaming body panels without welds; it found the method satisfactory where stresses on the joints were moderate. New riveting methods as well as adhesives are increasingly used where nonferrous materials, such as aluminum and composites, must be joined.

Spot-welding made feasible for the first time the introduction of fully integral body structures for volume-produced cars. The concept of a unitized body and frame was patented on March 28, 1919, by Vincenzo Lancia, who was inspired by the structure of a ship. In 1922, Lancia introduced his Lambda model, an open tourer whose steel side and floor panels were fully stressed and offered 10 times the torsional stiffness of comparable body/frame cars. To facilitate his construction of suitable bodies (for Lancia was not a volume producer), he later adopted a platform frame for the Lambda. In 1937, however, his company began producing the Aprilia, which had a fully unitized body and frame.

Meanwhile, Ledwinka of Budd filed a 1927 patent on a type of unit construction and subsequently built an experimental car unifying the body with the frame's deep chassis sills. In the 1930s in the United States, both Chrysler, with the Airflow, and Lincoln, with the Zephyr designed and

The 1936–1937 Cord had a unit body/frame construction.

built by Briggs, following principles introduced by John Tjaarda, built cars with integrated frames and bodies. The Cord 810 of 1936 was of integral construction as well.

These vehicles relied heavily on frame-like box-section members throughout their structures that did not use sheet metal as efficiently as possible. Closer kin to the modern unit-built vehicle was the Nash 600 introduced in 1941. This was the work of Budd engineer Ted Ulrich, who subsequently joined the Nash staff. In the post-war years, Nash and later American Motors were strong advocates of unit construction.

Meanwhile, in Europe the unit-body pace was set by the same team that had brought the steel body to the Old World, Citroën working with Budd. The famous *Traction Avant* Citroën of 1934 numbered a steel unitized body/frame among its advanced features.

In a lower-priced category in Europe, the first unit-body cars were introduced by General Motors. Its first such applications were on the Opel Olympia of 1935 in Germany and the Vauxhall H model of 1937 in the United Kingdom. Similar to Lancia a decade earlier, the designers of these cars realized that more rigid body/frame units were needed to cope with the

Nash had been a strong advocate of unitized construction since 1941. Shown is the Nash Statesman Super.

First built in 1934, the front-drive Citroën was a pioneering unit-built car.

The 1935 Opel Olympia introduced full body/frame unitized construction to the popular car field.

independent front suspensions that were coming into use; both the GM cars used Dubonnet-type suspensions, and the front-driven Citroën had parallel wishbones.

The light and efficient Opel became the most studied car of its kind in Europe and was paid the compliment of being directly copied by Renault, whose Juvaquatre was launched at the 1937 Paris Salon. The Olympia evolved into the Kadett, which the Soviets liberated after World War II and transformed into the Moskvitch.

With only few exceptions, the passenger cars produced in the United States and Europe and later in Japan adopted chassis-less construction and thus relied fully on the body itself to provide structural strength. In the late 1950s, Ford's Lincoln was an early convert to unitized construction among America's largest cars, although some of its models later reverted to separate frames. Large Ford, Lincoln, and Mercury models continue in the mid-1990s with separate frames, as do GM's biggest cars in the Cadillac, Buick, and Chevrolet ranges.

Problems of corrosion and NVH introduced with all-steel bodies were relaunched with the adoption of the unitized body/frame. Noise was more easily transmitted from the road and drivetrain, and rust could lead to

The United Kingdom's first unit body/frame car, a Vauxhall, was first built in 1937.

deterioration of the main vehicle structure. The wide adoption of galvanized steel and other protective coatings has coped with the latter problem. Porsche was the first producer to provide its cars with fully galvanized bodies, starting in 1975.

A popular "halfway house" for vehicle builders not ready to commit to integral construction was and continues to be the use of a platform frame. This is a relatively shallow platform that performs the function of a frame and often integrates body parts such as the cowl structure and the wheelhouses. Even after adopting integral construction, many producers, especially in Italy, continued to make platform-framed vehicles to suit the needs of independent coachbuilders.

Examples of cars with platform frames are the VW Type 1 or "Beetle," the Citroën 2CV, the Borgward, and the Mercedes-Benz 180 and 220 of the 1950s. In all such instances, the attachment of the body to the frame was designed and intended to add substantially to the vehicle's overall stiffness. In the 1990s, platform frames remained in use chiefly for a few open-topped sports cars, for which they are well suited.

In 1951, the war in Korea led the U.S. government to limit civilian uses of some metals, among them nickel, zinc, and tin. Steel was not immediately threatened, but it was obviously a good time to consider the potential of alternate materials. One such was known as glass-reinforced plastic, or GRP. During World War II, radar domes and naval minesweeper hulls were made from this new material, a combination of polyester resins with woven fibers of fine glass, the latter also known as "fiberglass," which became a generic term for the finished material.

As early as 1944, Owens-Corning was cooperating with Henry Kaiser in the experimental development of GRP bodies for automobiles. Kaiser designer Howard Darrin built his first GRP car, a pontoon-fendered con-vertible, in 1946. That same year saw Owens-Corning helping engineer William Stout design and build a complete GRP body for one of his rear-engined Scarab prototypes. As a result of the Kaiser experiments and the initiative of Howard Darrin, a GRP-bodied sports car was produced by Kaiser on the Henry J chassis. After the manufacture of 62 prototypes, production of the Kaiser-Darrin started in December 1953. In all, 435 were produced.

Meanwhile Chevrolet had shown its GRP-bodied Corvette sports car as a Motorama "dream car." "The body on the show model was made of rein-forced plastic purely as an expedient to get the job built quickly," reported Chevrolet's chief body engineer, Ellis J. Premo, to the SAE. "But people wanted delivery right now, and so our management decided to build 300 Corvettes with plastic bodies during 1953, starting in June."

Matched-metal-die tooling was laid down to build this pioneering GRP body. Each body was composed of 30 parts considered major and 32 minor fiberglass pieces. "The total weight of the Corvette reinforced plas-tic body parts is 154 kg [340 lb]," said Premo, "the basic materials making up this total being 62 kg [137 lb] of fiberglass, 69 kg [152 lb] of polyester resin, and 23 kg [51 lb] of inert filler." The complete body, ready to be mated to the chassis at 11 points, weighed 186 kg [410 lb].

On the basis of GM's tests, engineer Maurice Olley reported that "the physical properties of fiberglass vary considerably with the percentage of glass in the mixture, the texture of the glass, and the hardness of the resin. Too hard a material containing too high a percentage of glass is likely to suffer from flex cracking or stone bruises in fenders and body panels. An acceptable material will have physical properties which are remarkably similar to those of the woods used in older body construction.

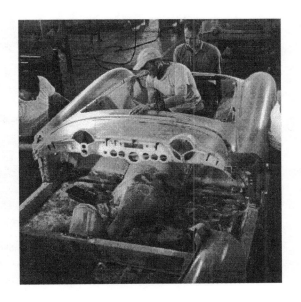

Building the GRP body of the original 1953 Corvette.

"What we get from all this," the British-born Olley summed up, "is a very usable body, somewhat expensive, costing a little less than a dollar a pound, but of light weight, able to stand up to abuse, which will not rust, will not crumple in collision, will take a paint finish, and is relatively free from drumming noise. A fiberglass panel of body quality three times as thick as steel will weigh half as much and will have approximately equal stiffness."

These were convincing arguments both in the mid-1950s and today. More than 40 years later, the plastic body introduced on the Corvette as an expedient continues to be offered, now with radically changed composition and production methods. The Corvette's new American rival, the Dodge Viper, is similarly bodied, as are many other sports cars.

Composite exterior and structural materials for car bodies must now be given serious consideration for any vehicle produced in volumes of up to 50,000 per year. Such cars as the Pontiac Fiero and Saturn show that the

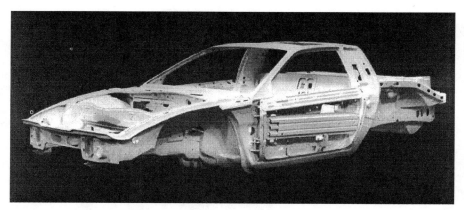

The space frame of the Pontiac Fiero.

potential can be even higher for the use of plastic exterior panels. In the MPV field, composite bodies are used successfully by the Matra-built Renault Espace and the GM range including the Pontiac Trans Sport.

Polymers have also proven useful for many other exterior body surfaces including impact areas. The first such application was in 1968 by Pontiac, which fitted an Endura bumper to its GTO model. This had a tough body-colored plastic surface on a steel core. Variations on this theme are now widely seen throughout the world motor industry.

Aluminum, as well, is challenging steel as a body material in the 1990s. This marks a comeback from the early years of the century when aluminum was widely used for that purpose. In the early 1950s, Panhard showed

Saturn adventurously adopted exterior composite panels on a steel structure.

Pontiac Trans Sport: a modern application of composite skin panels.

*The all-aluminum body of the 1954 Panhard Dyna
offered striking economy of operation.*

With its all-aluminum body, the Audi A8 sought a new way to build a luxury car for the 1990s.

what aluminum could achieve with its four-door Dyna model. Its body weighed 115 kg (253 lb); made in steel, it would have weighed 227 kg (500 lb). Although costing a customer 15.4% more than a steel-bodied version, the total operating cost of the aluminum Dyna, including depreciation, was 20.2% less over three years.

Similar reasoning underpinned the investment of $650 million made by the VW Group's Audi in aluminum body technology, in cooperation with Alcoa. Using a unique "space-frame" technology created to take full advantage of aluminum's properties, Audi launched its A8 model in 1994. Other auto-makers are actively exploring aluminum's potential for bodies, e.g., Ford with Alcan.

Since the 1950s, aluminum in anodized and polished form has been widely used as an exterior trim material. Plated plastics and stainless steel have been similarly employed. A complete brushed-stainless (austenitic SAE 304) exterior skin was used by the DeLorean sports car of 1981. Inspired by Allegheny Ludlum's all-stainless 1936 Ford and 1967 Lincoln prototypes, the DeLorean's GRP body was clad with panels pressed by Läpple from BSC Stainless brushed sheet.

Another feature of the DeLorean was its use of upward-hinged "gullwing" doors. Such doors were used on sports-racing cars by Mercedes-Benz in 1952 and then on the production 300SL in 1954. The short-lived Bricklin

The DeLorean was notable for its stainless steel skin and gullwing doors.

sports car of 1975 also used such doors, which represent the principal alternative to conventional hinged doors adopted in production by auto-makers.

If minivans are included, their sliding doors constitute another important and practical alternative. In spite of their successful use on many convincing prototypes, such as Bertone's Ramarro, sliding doors have failed to make progress on conventional auto bodies.

Other alternatives to the conventional door include the swing-up doors of the Lamborghini Countach and Diablo and also of the Bugatti EB110. Doors opening at the front of the car to permit driver and passenger to exit forward were shown in the 1930s by Dubonnet on a prototype. The concept was used in the 1950s by the Isetta, first by Iso and later by BMW, and also by Heinkel.

At the rear of the car body, luggage was initially carried in a separate trunk. In the United States, this continued to be the name of a luggage compartment when it was integrated into the design of the body, as occurred during the 1930s.

However, the traditional concept of a separate trunk continued to represent the more expensive and elegant design solution. Even in the 1990s, a separate trunk or a replica of one—or a hint of the presence of an outside-mounted spare wheel, also absorbed into the trunk—is a feature of modern cars built to resemble those of the great classic era.

Meanwhile the designers of volume-produced cars grappled with the problem of improving the flexibility of stowage within the vehicle to meet the changing needs of the family auto. Kaiser tackled the challenge with the large rear hatch of its Traveler model, the first of which was introduced as a 1949 model. A breakthrough vehicle was the Renault R16 of 1965, with its rear hatch for easy access and movable seats to convert passenger space to luggage capacity.

This concept, which spawned the versatile and practical "hatchback" auto body, severely tested the skills of body engineers. Previously they had been able to rely on a welded-in bulkhead behind the rear seats to stiffen the rear of the structure; now for hatchback cars such as the VW Golf (1974) and Plymouth/Simca Horizon (1978), they had to engineer ingeniously around the frame of the rear hatch to replace the stiffness lost with the removal of the panel.

Before World War I, German body designers were already beginning to make a transition to an improved integration of lines and shape of the hood with the main body of the car. This was a result of the rapid aerodynamic evolution forced by the Prince Henry Trials, for which the major companies built special open touring cars.

After the war, the rapid development of aeronautics caused designers to look anew at the shape and function of the auto body. Even compared to the crude aircraft of the day, their shapes seemed to be battling through the air. This inspired both Edmund Rumpler and Paul Jaray to develop aerodynamic auto forms. When tested in a modern wind tunnel, Rumpler's limited-production *Tropfen-Auto* of 1921 was shown to have a drag coefficient of only 0.28.

While Rumpler's designs inspired Benz and later Mercedes-Benz racing cars, Jaray's patents were acknowledged by Chrysler, which paid him a small fee for the Airflow. In the 1930s, Erwin Komenda of the Porsche design office and Hans Ledwinka at Tatra advanced the art and science of the more aerodynamic car. Principles first elucidated then were redis-covered in the 1970s and 1980s to create more economical cars for a fuel-straitened era.

Characterizing all the efforts of the aerodynamic pioneers, beginning be-fore World War I, was the use of curved windshields. This was an inherent element of the well-streamlined car. In the 1930s, a few Adler sports coupes were produced with curved windshields. Likewise, before World War II the Studebaker Skyway coupe used a flat windshield bent at the center, a technique the company continued after the war.

The watershed for the introduction of curved windshields in production cars was the late 1940s. The year 1948 saw them in the radical new Hudson, the Cadillac range, and its sister, the Oldsmobile 98. The center pillars were eliminated in the curved windshields introduced in 1949 by Nash (Airflyte) and Lincoln (Cosmopolitan). GM adopted one-piece curved windshields widely in its volume ranges in 1950.

The 1948 Cadillac Series 62 sedan had the first curved windshield in a series-produced car.

Next came curvature of the side glass as well to help the stylists enhance the "fuselage" look of the automobile. It was used by Chrysler on its 1957 Imperial and with great distinction and elegance on the 1961–1963 Lincoln Continental and Ford Thunderbird. By 1965, curved side glass had been widely adopted throughout the U.S. auto industry.

The 1983 Audi 100 was the first production car to achieve fully flush side glass, with the conjoint objectives of improved aerodynamics, reduced wind noise, and a new vehicle appearance. In the 1990s, Mercedes-Benz introduced double side glazing on its S-Class to improve insulation and reduce noise transmission.

Laminated safety glass was first shown as early as 1906, as were other concepts such as wire embedded in the glass. By 1910, laminated glass became practical, and its use was gradually extended in windshields. Ford's adoption of Triplex safety glass in the Model A marked a break-through in its application to the everyday auto. Toughened-glass windshields became and remain the preferred safety-glass technique in Europe. Around 1916, the first windshield wipers were introduced.

At the beginning of a new century of car design, engineers are experiment-ing with the new materials mentioned above as well as other techniques to make tomorrow's car body even more efficient in all respects. For them, intensifying preoccupations are the requirements of eventual scrapping and recycling of auto bodies as well as the unrelenting pressure to reduce vehicle weight.

Even thinner panels are possible through the use of bake-hardening steels that are easier to form before hardening. With tailored steel blanks, the thickness can be adjusted to meet the local stress requirement. Improved computer simulation will sharply reduce the physical-testing demands on both the aerodynamicist and the crash-safety engineer.

The pioneers of the car's first century have made great progress in all as-pects of body design and construction. They have, in fact, placed the body more at the heart of the automobile. Having done so, body engineers and designers today face new and exciting challenges. The pioneers have not had all the fun.

A Century of Automobile Comfort and Convenience

Karl E. Ludvigsen
Photos courtesy of The Ludvigsen Library Ltd.

In the early days of the automobile, the driver and the passengers adjusted themselves to the car. Now the car adjusts itself to them in the most comprehensive and remarkable ways. In this, the use of power-operated systems and the application of ergonomics or human factors combine to show one of the main ways in which car comfort has advanced over a century.

Climate control also has advanced, from the early days of the open touring car to the modern enclosed car with its sophisticated HVAC system, including filters to cleanse the incoming air. No longer do we need to open the windshield to enjoy a cool car interior.

Motoring itself provided more than enough entertainment and adventure in the early days. When available, however, radio became an obviously attractive accompaniment to auto travel. Radio will also play a key role, in new ways, in the traffic guidance and navigation systems that will secure the automobile's position as a means of mobility in decades to come.

1896 to 1914:   Car Comfort to World War I

Throughout these years, the opportunity to ride in the open in fine weather through the countryside was an important part of the pleasure of owning an automobile. Special clothing protected owners and their families from the dust and mud of the early roadways.

The most important comfort element added to cars in this period was the windshield. The use of windshields in early cars was a matter of some difficulty. There was no easy way to keep them clean, for example. They were easily shattered by stones from the rough roads of the period. In addition, the bodies of the earliest cars offered no place to mount a windshield where it would be of significant value.

The French led the way in the design and construction of closed auto bodies—an obvious advance in comfort. In 1903, the first fully enclosed American car body was built by Duryea.

Side windows, when introduced, were initially fixed in position. Next they were made to slide up and down similar to the windows of a railway car and held in place by a strap with a number of holes used to adjust the height of the window.

Seating in the earliest cars resembled that of horse-drawn vehicles. Bench-type seats were upholstered in tufted leather, resembling a Chesterfield sofa. No other material or design would suit the requirements of the early open automobile.

After closed cars were introduced, it became possible to employ fabrics for seating. Thus began the long and continuing process of developing specialized fabrics well suited to automobile applications.

During virtually all of this period, the automobile retained a traditional dashboard, a vertical panel set directly behind the engine hood. Forming the front of the passenger compartment, it literally was a board against which pebbles were dashed from the roadway. Usually the windshield, if fitted, was mounted directly on the top of the dashboard.

This dashboard was the only surface to which instruments could be attached. Thus any instruments were located very low, slightly above the driver's feet—not ideally placed for viewing. On the other hand, they did not require frequent surveillance. They could also be placed close to the mechanisms they were designed to monitor.

Flooring made of wood was normally used. In the driver's compartment the flooring was often covered by a thin sheet of patterned metal, such as aluminum. This preserved the driver's footing even when the weather was bad and the floor was running with water.

In the first closed cars, the floor of the rear compartment was covered by cut-pile carpeting similar to that used in residential interiors. Carpeting was a clear indication of a costly luxury model because it was not provided on volume cars.

The pioneering open cars had few places to store personal items taken aboard, exposed as they were to the wind and the elements. When doors gradually were fitted, late in the period, they were provided also with door pockets. These were the earliest forms of interior stowage, particularly made available on luxury models.

In the early open cars, comfort was more a matter of shielding the occupants than heating or cooling them. When weather protection was erected, however, it was necessary to provide some means of admitting fresh air to the interior on a controllable basis. For this, a portion of the windshield could be opened to the degree desired.

One of the first cars to provide purpose-designed interior heating was the 1910 Delaunay-Belleville supplied to Tzar Nicholas of Russia. Its Kellner limousine body had a heating system for the rear compartment using hot water from the engine.

Not until the end of this period were the first complete electrical systems installed. Beginning in 1912, Cadillac in America was one of the first to make available a complete electrical system incorporating electric starting.

Magnificent oriental-style interior of the 1920s British Sheffield Simplex with custom body. Note the dome light and locks to hold the sliding windows in place.

At an earlier date, top-level cars were equipped with interior electric lighting. One of the first with this feature was a Daimler bodied by Hooper in 1904 for the British Royal Family. The car was equipped with full electric lighting in the rear passenger compartment.

As soon as electrical systems were provided, interior lighting became standard. This usually took the form of a dome light for the rear passenger compartment.

1918 to 1933: The Maturing of the Automobile

A significant advance after World War I was the introduction of crank-operated roll-up windows. Various types of mechanisms were used; in some cases, the window glass was spring-counterbalanced so that it could be raised or lowered quite easily.

Two early car interiors: the upper custom body has sliding side windows, and the lower one has early crank-operated windows. The footrest in the upper car is movable.

Interior of the 1932 Cadillac V-8 six-window sedan had wood trim and pull-down shades for rear and quarter windows.

On most cars through this period, the windshield remained vertical. It could normally be opened, usually from the base and hinged at the top, to provide interior ventilation. In addition, for luxury models vents were introduced at the sides of the cowl to admit cooling air to the front footwell. On sporting cars, the windshield was built to fold flat when the top was down.

The better class of closed car had roll-down blinds or shades for all windows. Resembling those in railway cars or private homes, they could be pulled down into position for privacy or sun protection. In some bodies, the driver or chauffeur could remotely draw down the rear-window blind as a shield from the headlight glare of following cars.

In the custom bodies built during this period, seats often were not adjustable. Adjustability of seats was introduced during this period but more often for volume cars than for luxury models, which were custom built to suit the dimensions of their drivers and passengers.

Seat construction used a base of coil springs covered by a horsehair-filled cushion below the upholstered surface. For the seats, the objective was to have cushioning supple enough to provide a satisfyingly soft sensation but at the same time able to prevent the passenger from "bottoming" when the car hit a bump in the road.

Coachbuilders had access to a variety of sizes and types of springs and cushions which they used to tailor the seats of a custom-built car exactly to the tastes of the purchasers. As much as 0.2 m (0.66 ft) of spring depth was available to the seat designers. Sometimes they used two layers of springs, a lower layer made of heavy wire and a softer upper layer of fine wire.

The seating of fine luxury cars tended more and more to resemble the furniture their owners would use at home. The model for rear-seat design was a sofa with side arm rests. The sofa similarity was enhanced in many designs by the use of extra loose cushions which added to both the appearance and feel of luxury.

Interior heating of luxury cars concentrated on the requirements of the rear-seat passengers in town cars and limousines. For them, hot-water heating systems were provided in the most expensive luxury cars in both Europe and North America. However, the driver, often exposed to the elements, was not considered to need heating.

For practical reasons, the custom body builders favored leather trim for open cars and fabrics for closed cars. They took pride in designing open cars so that they could be used as such in all but the worst weather conditions. Thus, they usually had highly serviceable leather seat trim.

The earlier style of tufted upholstery was largely abandoned during the 1920s. Fabric trim styles became more restrained, with pleats or simple upholstery designs, resembling the fine furniture of the period. Both leather and pigskin were used for seat trim. Leather sources were carefully screened to be sure that the cattle were from areas where barbed wire would not cause scratches.

For fabric, wool broadcloth was favored. Bedford cord fabrics also were used. Manufacturers made these fabrics available in a range of colors and patterns. Finely striped fabrics were used as well. Custom coachbuilders kept books of fabric samples in their offices from which customers could select those that they preferred.

Interior trim items such as robe rails and assist straps were often beautifully woven of fine fabrics in a braided pattern. A custom body would be supplied with a lap robe to match the interior and with extra cushions for the rear seats, complementing the rear seat trim. Plain fabrics were used for headliners.

Toward the end of this period, the first synthetic fabric materials began to be used in custom bodies. Rayon was the first available. Woven to form abstract patterns, it was used in custom bodies as early as 1928.

Late in the 1920s, the first radio installations were made, installed at first only in rear compartments to avoid distraction of the driver. This was an ideal novelty to appeal to the owners of luxury cars, especially those with custom-built car bodies.

Radios were first offered as options on American cars in 1929. In 1930, Chryslers were fitted as standard with all the necessary wiring for a radio installation. Most cars still had fabric roofs, supported by a chicken-wire structure. Insulated from the main body, this chicken wire was used as the radio antenna.

At the end of this period, switches were introduced to automatically turn on the interior lights when the doors were opened. Initially this system was used for the rear compartments of custom-built cars.

Some custom-built cars were fitted with lights which illuminated the step or running board that the owner used when entering or leaving the automobile. Literally a "courtesy" light, this was very useful for the high-built cars of the era.

1934 to 1941: Dramatic Changes in Car Architecture

An important advance during the 1930s was the introduction of small front quarter-windows that could be opened to admit air without the window itself having to be wound down. GM's Fisher Body introduced this on all 1933 GM-built models as "No-Draft Ventilation," and Packard also offered the feature. From 1933 to 1935, most auto-makers introduced some variation of this idea.

At the same time, more producers in both Europe and the United States were introducing opening air vents in the cowl. Air-pressure distribution made this cowl-fed ventilation highly efficient; some producers had offered it in the 1920s, and it was widely used by the mid-1930s. Combined with quarter-windows, the cowl vents allowed windshields to be fixed in place, which in turn accelerated the development of vee-shaped and later curved windshields.

Seats became universally adjustable. The European arrangement of individual front seats enabled the driver and passenger to adopt different positions according to their respective sizes and desires. Also typically European was adjustment of the angle of the front seat back.

Adjustment also began to be provided for the rear seats of luxury cars. On Fleetwood-bodied Cadillacs, for example, both the depth of the cushion and the angle of the seat back could be adjusted in the rear.

Fold-down center arm rests were widely introduced. This was most common in bench rear seats, but was also used in bench front seat designs such as the Cord 810. Arm rests were also fitted at the sides; on a luxury car, those for the front seats were mounted on the doors and were adjustable for height.

The Chrysler Airflow and Lincoln Zephyr broke away from the traditional furniture-style look of automobile seating. Both were inspired instead by the seats being used in aircraft, with their exposed tubular structural members. The more modern styles in furniture being developed by Marcel Breuer and others also influenced these designs.

New developments helped clear the front floors of many prestige models. Late in the 1930s, the handbrake lever was moved away from the floor and became a suspended lever under the instrument panel. The shift lever was moved to the steering column where it was easily reached and manipulated. By 1938, both of these features had been incorporated in Cadillacs.

American luxury cars began to make provisions for heaters. Heater cores and fans were, in many cases, mounted beneath the front seats where they could provide warmth to both the front and rear compartments. In Europe, heaters continued to be regarded as an option, even for luxury cars, and were not integrated in any way into vehicle design.

At the end of this period, the first air conditioning systems were introduced. Packard was first to offer air cooling in 1940. Cadillac followed in 1941.

The interior of King Edward VIII's 1936 Buick Limousine had drink and refreshment cabinets, a folding light, mirror, pencils, and a microphone to communicate with the chauffeur.

Leather trim, previously seen as chiefly appropriate for open cars, began increasingly to be used for closed cars as well. Thanks to its open-car associations, leather had and continues to have a sporty connotation.

The first serious uses of plastics in car interiors characterized this period. In fact, plastics had been introduced at the end of the previous period. In 1931, for example, Le Baron used plastic knobs for the interior hardware of a special Lincoln limousine. The knobs were cut and machined from a rod of nitro-cellulose plastic.

Late in the 1930s, Graham and Hupmobile used plastics for instrument panel components. Henry Ford was an ardent advocate of soybean-derived plastics, using them for steering-wheel rims, shift knobs, and horn buttons. The early plastic parts used in these applications lacked durability, however, especially when used in parts that were exposed to the heat of sunlight.

1941 Cadillac Series 62 Convertible coupe with column shift to clear the floor area.

A significant development was the provision of the usable glove compartment. This made available a large lockable compartment for various items carried onboard. Previously, small compartments had been provided at the sides of the centrally placed instrument panel. On volume cars, these might be open cubbyholes or simply a package shelf beneath the panel.

Automatic interior lighting, from a central dome light, became well established in volume-built and custom luxury cars. A separate switch, usually on the B-pillar, was provided to switch on the light when desired.

The ambitious Auburn group was a leader in integrating radios into automobiles. The 1935 Auburn Speedster was the first American car with a radio fitted as standard with front-seat controls. In 1936, the Cord 810 (also built by Auburn) offered a radio fully integrated with the design of the instrument panel.

In 1935, a special Duesenberg bodied by Bohman and Schwartz was equipped with a built-in radio, also with front-seat controls. Separate rear-seat controls for radios were continued for limousines, located in more convenient positions such as the passenger arm rests.

When all-steel bodies were introduced, it was no longer possible to use the chicken-wire roof support as an antenna. Antennae were then mounted under the running boards. They, in turn, had a short lifetime, and the whip antenna had to be introduced.

In 1939 and 1940, American cars were the first to offer an important feature of the automobile radio: pushbuttons to select predetermined stations. For the convenience of the driver, naturally busy with other concerns much of the time, this was a very important and useful advance.

1945 to 1956: A Decade of Postwar Progress

Open cars during this period enjoyed the wide application of power operation for their canvas tops, an amenity first offered on a production car by Chrysler's 1939 Plymouth. Electric and/or electrohydraulic mechanisms were increasingly used to raise and lower convertible tops.

Another important innovation during these years was the introduction of power-operated side windows. Power operation was introduced for front-seat adjustment as well. A switch at the side of the seat controlled the adjustment. Initially the power adjustment was fore-and-aft only; toward the end of this period, an adjustment for height was added.

An important distinction of European luxury-car seating during this period, differentiating it from both American luxury cars and volume cars, was the continued provision of front-seat back-angle adjustment. American car designers were slow to recognize the appeal and attractiveness of this feature.

An important European convenience feature was the use of a steering wheel adjustable for its distance from the driver. Jaguar was among those providing such adjustment, a feature perceived as being highly desirable by luxury-car purchasers.

While Europeans in general, and the British in particular, favored wood-paneled trim, the American car-makers quickly learned how to simulate wood grain on metal surfaces. They began to use this technique extensively in their interiors in the 1930s and continued to deploy it widely in this period. Cadillac's handsome 1948 panel was an excellent example.

The attention given to the functional design of products for World War II greatly accelerated the science of ergonomics or human factors. From measurements taken of people enlisting in the armed forces, data were available for the first time of the measurements of a substantial sample of the male and female population of the United States. This provided the basis for the first serious effort to design products, including automobiles, to better suit people.

Heating systems became integrated fully with car body design. Instead of being add-on units, often encumbering the front compartment, heaters were built in. This also made possible the provision of fresh heated air, drawn from the front of the car. At the same time, heater controls became integrated with the instrument panel instead of being attached to the body of the heater.

Associated with this integration of the heater, the provision of fresh air for ventilation through ducts from the front of the car also became more common. In this respect, American cars substantially led their European counterparts, which were slower to adopt more modern ventilation systems.

European luxury cars introduced heating and ventilating systems that allowed different temperatures to be maintained on the left and right sides of the front compartments. This became a distinctive luxury car feature on European models such as Mercedes-Benz.

Air conditioning was confirmed during these years as an important optional feature for luxury cars. System technology advanced rapidly. The provision of fresh and recirculated refrigerated air was made possible.

Custom body building continued to be practiced in great style, especially in Britain. A 1952 Daimler bodied by Hooper had lavender blue leather upholstery with lizard skin trim borders. The Daimler was comprehensively equipped with cabinets holding a full range of picnic and cocktail accessories.

A clear differentiation between European and American stowage practice became evident in this period. Apart from the glove compartment, American cars ceased providing any other interior storage capability. In contrast, European models continued to offer door pockets and other stowage provisions in addition to a glove compartment.

American car designers moved quickly in this period to fully integrate radio units and speakers with the design of instrument panels. Their counterparts in Europe, however, were reluctant to accept the fact that the radio could be more than an afterthought, suspended below the instrument panel as an accessory rather than a factory option.

In addition to the dome light, cars increasingly offered downlights placed under the cowl to illuminate the front floor when the doors were opened. This provided much better general interior lighting.

European cars, especially those built in Britain, often provided a map-reading light for the passenger. This was and remains an important distinctive lighting feature. It was neatly incorporated into the panels of some American cars such as Buicks.

1957 to 1967: New Automotive Influences

Accelerated pressure for innovation during this period produced new ideas for entry and exit. Some American cars had separate front seats that swiveled outward, easing entry and exit, when the front doors were opened. Chrysler in particular offered this feature.

General Motors introduced an adjustable steering wheel, designed and built by its Saginaw Division. A single pull of the control lever let the wheel move to its highest position, so that entry and exit were made easier. In addition, Cadillac offered an adjustment for reach, unlocked by a twist control on the wheel hub.

The Ford Motor Company introduced its "swing-away" steering wheel. When the automatic transmission was placed in "Park," the swing-away wheel pivoted on its column to the right, toward the center of the car, easing exit and subsequent entry. Re-entering the car, the driver had to swing the wheel back into position before the transmission selector could be moved.

The rediscovery of individual front seats for many cars, especially in the United States, was an important feature of this period. Two of the most significant cars were the Ford Thunderbird of 1958 and the Buick Riviera of 1963. Both had completely individual front seats divided by a deep center console.

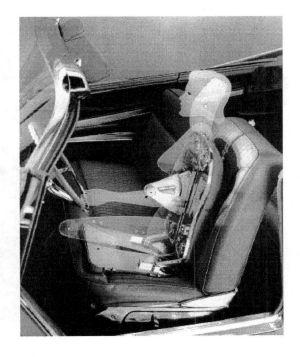

Car companies each had their own ergonomic measuring devices before an SAE standard was adopted.

From 1957, Mercury and Ford's Thunderbird offered a seat-position re-trieval system called "Seat-O-Matic." After a driver had adjusted the seat to his/her taste from the five vertical and seven longitudinal positions of-fered, the position corresponded to a letter-number combination that could be dialed-in to retrieve the setting. When the ignition was turned off, the seat went all the way back and down to offer more entry/exit room.

Advances were rapid in the application of ergonomics to interior design. In the early 1960s, General Motors set up a Human Factors Unit that con-ducted work of a high standard. Later in the decade and through the early 1970s, the study and application of ergonomics advanced in the United Kingdom, Sweden, and France.

The ASTRA-DOME instrument panel in 1960 Chryslers was lit by seven wafer-thin electroluminescent panels, and its center, which was said to resemble a planet in space, grouped all controls, gauges, and dials around it, directly in front of the driver.

Cadillac achieved a breakthrough in climate-control equipment. In 1964, it began offering an optional air conditioning/heating system in which the driver had only to select the interior temperature that the system was to maintain. Using sensors in the car interior and in various system ducts, the climate-control system electronically adjusted air conditioning, heating, or both to provide the required interior temperature level.

The major advantage of this system was that it greatly simplified instrument panel controls. These had been becoming very complex to monitor and control all aspects of heating and air conditioning. On a Rolls-Royce, for example, it was possible for the driver to select up to 1,380 different permutations of the various heating and refrigeration controls.

At the end of this period, American manufacturers adopted air conditioning as standard equipment on their top-of-the-line models. In limousines, rear-seat passengers were provided with their own controls to regulate compartment temperature.

Important advances in radios in particular, and in-car entertainment in general, occurred during this period. In Europe, receivers were typically multi-band design to take advantage of all broadcasting. In the United States, radios increasingly were able to receive FM and AM, as the FM broadcasting network developed and expanded.

In the United States, the first efforts were made to play pre-recorded music in automobiles. Chrysler's "Highway Hi-Fi" of 1956, for example, used special 16 2/3-rpm records, and others used turntables playing the small 45-rpm records of the time. These were not up to the demands of bumpy-road operation, however. Instead, the first successful car-borne systems used so-called eight-track stereo cassettes. Although bulky, these cassettes worked well in automotive applications and achieved wide acceptance.

These years also saw the widespread application of signal-seeking capability to car radios. This practical advance, which had been pioneered in 1950 by Delco, allowed the driver to search for suitable stations with less distraction. Some cars placed the control for signal-seeking on the floor so the driver could manipulate it with his foot.

Another advance was the availability, particularly in America, of multiple-speaker stereo sound systems. Improved understanding of the interior acoustics of the body made these systems extremely effective. Dramatic improvements in sound quality, especially for music, were achieved. Sound systems in automobiles began to rival the progress being made in high-fidelity systems for the home.

Interior lighting moved forward with the introduction of several new configurations. In some GM designs, a general interior light was combined with one or two small aircraft-style spotlights that could be moved to direct light on a map without distracting the driver.

Switches to control interior lights were combined with the light itself, typically in Europe, or controlled by means of the headlamp switch, usually the case in the United States.

1968 to 1980: Interior Safety—A New Challenge

Developments in seat design in the 1970s were conditioned by one powerful influence: the pressure to improve car safety. Head rests, which had been fitted from time to time as a luxury feature, were now required by safety regulations to resist whiplash. Head rests were at first fitted to the front seats of all cars marketed in the United States and later in the period to rear seats as well.

American cars finally adopted various adjustment methods for the front-seat back angle. In addition, six-way power seat adjustments were introduced. These adjusted the overall angle of the complete seat, in addition to its longitudinal position and height. Six-way adjustment became an important luxury-car feature which was relatively slow to filter down to volume cars, even as an option.

Toward the end of this period, the seat-adjustment controls were mounted on the door rather than on the seat. This made the controls much more visible and accessible, rather than leaving them concealed at the side of the seat.

The 1976 Cadillac Seville had electric window controls on the door, folding front arm rests, and a "split bench" seat.

European luxury cars began to pay more attention to the need for efficient air conditioning and heating. In 1969, air refrigeration (as it was called) was made standard equipment for the first time by Rolls-Royce. German and Italian car builders also began to integrate air conditioning more fully into their vehicle designs.

Cool-air ventilation was substantially improved by the use of through-flow systems. These were designed in such a way that the air drawn in at the front was easily exhausted through vents or specially designed door seals at the rear of the car. Improved knowledge of aerodynamics and exterior pressure distribution allowed such systems to function much more efficiently.

Both fresh- and refrigerated-air systems demanded larger and more numerous ventilation outlets in the instrument panel. The provision of these outlets and their adjustment became a major headache for the stylists and engineers.

Automatic climate-control systems, such as this example from the 1978 Buick Electra Park Avenue, greatly simplified HVAC controls.

Stowage arrangements suffered during the period because instrument panels and interiors were being redesigned for greater safety. Designers tried to compensate with larger door pockets and increased use of the console for stowage, especially of tape cassettes and other small items.

The Philips system, using cassettes much smaller than the eight-track type, gradually became established as the norm for passenger car in-car entertainment systems. This was a big improvement over the previous system because the cassettes occupied much less space. Cars also began to be equipped with compartments in which cassettes could easily be carried.

In the United States, Citizen's Band (CB) radio enjoyed a spectacular boom in popularity. CB sets were fitted to cars as accessories. In addition, auto producers began offering CB equipment as factory options in units that also incorporated normal radio receivers.

An attractive and useful feature, introduced first on luxury cars, was the provision of lighting for the vanity mirror mounted in the sun visor. This was a practical and much-appreciated amenity, lighting automatically when the mirror was uncovered. Illuminated vanity mirrors were initially introduced in the United States and were soon adopted by European producers.

1981 to 1995: Luxury, Electronics, and Creativity

Adjustability for the rear seats of luxury cars, usually power-operated, became a distinctive feature. Power positioning of head rests was also introduced. A symbolic seat operating control was introduced in Europe, making it easy for occupants to visualize seat adjustment.

More effective use of foam cushioning for car seats was introduced. Pioneered as long ago as 1939 by Hudson in its Country Club series, using Goodyear techniques, foam cushioning made it possible to design more unusual seat contours. It was especially useful in giving cars more deeply bucketed front-seat configurations.

Leather, once the material of choice for open luxury cars, now became the most prominent seat-trim material in this category. Although fabric seating materials continue to be offered, particularly in the United States, the role of leather as the main luxury seating material became virtually unchallenged.

The "loose cushion" look was popular during the 1970s and 1980s in luxury models.

Various forms of steering wheel and column adjustability were taken for granted on cars of all types. The Porsche 928 added a new dimension by adjusting the complete instrument nacelle along with the steering-column height.

Fully automatic systems controlling both heating and cooling came into widespread use on luxury cars in both North America and Europe. This was facilitated by advances in electronics and servo systems.

Also confirmed during these years was the use of ducting that provided warm defrosting air to the side windows and the windshield. On some models, such as the Lincoln Continental Mark IV in 1974, the first applications were made of a new type of electrically heated windshield for defrosting.

Continued miniaturization of electronic systems allowed cassette players and radios to be combined in units of much smaller size than had hitherto been possible. This allowed manufacturers to offer, as options, much more powerful amplifiers with sophisticated frequency-adjustment capabilities.

Introduction of the compact disc for sound reproduction represented the next major advance in automotive in-car entertainment. Luxury-car-makers offered compact-disc systems of various types. Some allowed the discs to be changed in the passenger compartment; in others, a number of discs were stored in an automatic changer, with programming controlled by the driver or passenger.

The mobile telephone took the place of CB equipment for two-way communication. Car designs began to take into account the need to provide for the installation of telephone equipment.

In the 1990s, the first on-board navigation systems for luxury cars moved out of the prototype stage and into production. It is already evident that these systems will represent extremely important equipment items in the in-car entertainment and instrument categories for luxury cars. In spite of their obvious value for all drivers, their cost is likely to keep them for some years a differentiator between luxury and volume automobiles.

Lighting systems for car interiors were made more user friendly. In some cars, special lights were provided to show the position of the ignition switch and/or the door-lock keyhole. In others, interior lights were turned on when the door handle was moved or when the door was unlocked.

In addition, time-delay systems were used to keep interior lights on, for convenience, for a period of time after the driver's door was closed. Automatic "theater-light" dimming when the lights go off became a pleasing provision, typical of the thoughtfulness that has made the modern car a pleasure in which to ride and drive.

The First 100 Years of Transportation Safety: Part I

Anthony J. Yanik

Transportation safety has been defined in various ways over the years, each succeeding generation being faced with the task of improving on the efforts of predecessors. In the early days, transportation safety was considered the domain of the driver. It was assumed that automobile manufacturers had given the operator the appropriate tools for driving safely, such as reliable and durable brakes, steering, lighting, wheels, and suspension systems. It was up to the driver to use these tools in a manner that would safeguard the driver and passengers, as well as occupants in other vehicles, bicyclists, and pedestrians.

World War II inadvertently caused a change in this outlook inasmuch as the military, particularly with regard to safeguarding aircraft pilots, became more conscious of the need to tailor protection to the users of such complex weaponry. The aircraft ejection seat is a prime example of such thinking. Following World War II, engineers and researchers returning to public life felt that this novel approach toward safety design gradually could be transferred from the aircraft to such consumer products as the automobile.

Why "gradually"? First, the marketplace was not ready for such a shift in attitude toward vehicle safety. More importantly, even if the market had developed a groundswell of demand for occupant protection features at that earlier date, manufacturers would not have been in a good position to respond because they lacked elementary tools for safety development testing such as full-scale barrier crash sites and high-speed impact sled test facilities. Also lacking were reliable test dummies as the sophisticated devices necessary to record the dynamics and severities of test crashes.

Over the past four decades, however, the technology for improving automobile crash protection and crash avoidance has advanced remarkably. In response, the number of vehicle fatalities has decreased more than anyone would have been willing to predict in the 1950s, especially in the face of the growing numbers of vehicles on the road.

The Early Years

Mr. H.H. Bliss would have preferred to avoid the fate that awaited him when he stepped off the trolley car in New York City on September 13, 1899. At that moment, Arthur Smith, driving a new electric cab, decided to pass a slow wagon on his left by looping around the right side of the trolley just as Mr. Bliss stepped into his path. Mr. Bliss never heard what hit him as he became the first automobile fatality of which we have record. Mr. Smith, the cab driver, was not held, symptomatic of the attitude over who was at fault in traffic accidents, an attitude that was to prevail for many years. Mr. Bliss was transported to the cemetery in a horse-drawn hearse. Lest we lose perspective, there were 200 other fatalities on New York streets that year—all caused by horses.

Automobiles were not much of a bother on the roads at first, even with their rudimentary braking and steering apparatus. Speeds were so low—no more than 15 to 25 km/h (9 to 16 mi/h)—and motorcars were so few that collisions were infrequent and not often serious. Gradually, however, vehicle speeds began to increase, and with this increase came a

This 1890s crash resulted in passenger ejection and fatal head and neck injuries.

proliferation of wealthy "playboy" drivers who drove their high-powered vehicles with reckless abandon. In those early days, it was they who gave the automobile a bad name, especially among farmers.

Higher automobile speeds accelerated the development of better brakes and steering. By 1900, the more manageable round steering wheel replaced the tiller, and in 1902 Sterling Elliot invented the steering knuckle which enabled both front wheels to turn instead of the axle. Such developments made the negotiation of turns easier and safer.

Early brakes, however, were another challenge. Braking took place entirely on the rear wheels, and brake lockup was a way of life over the loose-dirt roads of the day. Drivers soon adopted the habit of steering their way out of accidents, if possible, because the stopping distances afforded by their brake systems were horrendous. In 1902, the Automobile Club of America measured the distance it would take to stop a motorcar at 32 km/h (20 mi/h) and found it to be an average of 18 m (59 ft)!

The Automobile: A Century of Progress

Little driving was done at night, one reason being the lack of adequate headlighting. Oil lamps were used with negligible effect until 1904 when Prest-O-Lite introduced a primitive spot-lamp system. It consisted of a steel cylinder filled with acetylene gas metered through a needle valve. The spot-lamp, once lit, may have been the equivalent of a 60-W bulb in a dark room, but it was the best available. The Guide company unveiled an electric lamp system in 1908, but effective lighting was not available until 1912 when Kettering introduced his generator/battery lighting and self-starting system.

By 1904, folding windshields of plate glass began to appear as optional equipment on some automobiles to protect their occupants from driving rain, bugs, and flying mud. But the windshields did not become popular until after 1910, probably because, by then, someone had the foresight to invent a hand-operated windshield wiper. Earlier windshields were split so that the top half could be tipped downward when vision was obscured by rain or snow. Drivers otherwise wore goggles.

Tires were a common source of irritation to early drivers. Several blow-outs on a single trip were not uncommon. More often than not, a car owner would carry two or three spare tires on a Sunday drive to avoid the possibilities of being stranded with the family too far from home. Typical of the time was the clincher tire and rim. The clincher tire was round and consisted of several layers of rubber-coated cloth having no tread. The clincher tire was very difficult to stretch over the wheel rim or to dismount. The demountable rim, introduced in 1906, made the task easier, and by 1908, tire-makers were beginning to experiment with an additional layer of rubber with indentations or tread to reduce skidding.

In the public sector, little was taking place relative to safety. A foreshadowing of things to come occurred during the 59th Congress of 1905 when a bill was introduced to regulate how vehicles operated on streets and highways, but it came to naught. In 1908, however, first Toledo, then Detroit, placed red/green semaphores in the center of key intersections, where they were operated by police.

The World War I Decade

During the second decade of the century, the proliferation of automobile companies that marked the first decade slowed considerably, especially when Henry Ford introduced the moving assembly line in 1913, with which small manufacturers could not compete. Probably the most significant safety advance of the decade was Charles Kettering's combination electric starting, ignition, and lighting system, first adopted by Cadillac in 1912. It was not the first starting system devised, but it was the most reliable for various reasons: its revolutionary motor/generator could apply short, powerful bursts of electricity to

C.F. "Boss" Kettering's original drawing (above right) for the patent for the automobile self-starter and ignition system which was introduced on the 1912 Cadillac. Above, Kettering is shown making adjustments during tests of the self-starter.

the starting motor without ruining the battery; it would feed electricity back to the battery after the vehicle was running, thereby keeping the battery charged; and it eliminated the onerous task of cranking the engine by hand, thus opening vehicle ownership to women.

Another portent of things to come was the development of the all-steel "open" body by Edward Budd, first used by 1912 Oakland and Hupmobile models. The all-steel open body was a revolutionary advancement over the typical wood body construction used for both open and closed cars of the day. Wood bodies were held together by wood screws and glue, and were prone to burst apart in a collision—especially if moisture had seeped within them, causing the screws to rust. Moreover, all-steel bodies allowed engineers to standardize body designs, and they were much more amenable to mass production. When the Dodge brothers adopted all-steel open bodies on their first cars in 1914, usage became more widespread.

The year 1912 was a boom year for safety. Also appearing that year was the first engine temperature gauge, the Boyce Moto-Meter. It was a common sight on vehicles, both as a radiator ornament and visual indicator of engine temperature to the driver. Marmon followed with the first rearview mirror, borrowed from the Wasp, which was the Marmon entry in the

This 1930s rollover test demonstrated the benefits of a car with a full-steel structural roof and occupant compartment.

Indianapolis 500 first held the previous year. Ray Harroun, who won that event for Marmon, used the mirror as a substitute for the riding mechanic, thus saving weight.

Other safety innovations of the World War I decade were the dash-mounted fuel gauge by Studebaker in 1914, the tilt-beam headlamp introduced by Cadillac in 1915, the push-button door lock located on the windowsills by Franklin in 1917, and the slanting of windshields to reduce glare in 1918. The first test track usually is considered to be the 800 m (875 yd) of wood road plus hill and sand pit constructed adjacent to the Dodge brothers' factory in 1915.

Traffic problems began to proliferate to the point at which communities stepped up with innovations of their own. Detroit helped pave the way, installing the first center lines in 1911 and the first stop sign in 1914. Cleveland introduced the red/green traffic light in 1914, paid for by the local street railway looking to avoid impediments to the flow of its rail cars. Finally, in 1919, William Potts, a Detroit police officer, developed the timed red–yellow–green light to control traffic in all four directions. That same year, the Detroit Traffic Division instituted the first school patrol.

At the federal level, Congress passed the Federal Road Aid Act of 1916, which provided for the building of roads within states according to a 50/50 matching formula. The government would select the routes and set the design formulas, and the states would provide for their upkeep.

The 1920s: Advent of the Closed Cars

The 1920s brought substantial progress in lighting, braking, tire technology, and windshield protection. By and large, the most significant trend that emerged was the shift of the buying public toward closed cars. Prior to the 1920s, open cars were the vehicles of choice for nine of ten Americans simply because they were so much cheaper than closed cars—by almost half. The Hudson Motor Company changed things in 1922 when it introduced the low-priced Essex Coach, a rather boxy, plain vehicle that nevertheless cost only $200 more than an open car. Its sales jumped by 1600% in six years, forcing other companies to bring out their own low-priced, closed sedans. By 1928, more than 80% of cars sold were closed versions. Essex had changed the entire market. Much credit also must go to Dodge, which introduced the first all-steel "closed" body (except for the roof) in 1923.

The popularity of the closed car brought with it concern for the safety of occupants from broken plate glass. In 1926, Stutz introduced a windshield through which ran horizontal wires every few inches to restrain loose glass pieces if a passenger struck the glass. The 1926 Rickenbacker went one step better, offering a laminated windshield in which transparent celluloid was sandwiched between two 3-mm-thick panes of glass. Ford introduced a laminated windshield on his new Model A in 1927, and in 1928 Cadillac added laminated glass all around.

Of course, windshields in closed cars were useless if they could not be kept clean during inclement weather. Thus, by 1923, the hand-operated wiper was being replaced by the vacuum-powered automatic type; an improved version, the electric wiper, was introduced by Delco Remy in 1925, although it took several years to catch on. Switching to inside vision aids, Studebaker introduced the built-in defroster vent in 1928.

Night driving was enhanced substantially when Guide pioneered the Tilt Ray headlamp in 1924, the first two-filament bulb. It allowed drivers to switch from low to high beams when circumstances permitted. GM moved the switch mechanism from the steering wheel to the floor in 1926. Earlier, in 1921, Wills St. Claire offered the first backup lamp that lit the road behind the car when the car's transmission was placed in reverse. It was combined with a stop light.

Although four-wheel mechanical brakes had been introduced by the Aland Motor Car Company in 1916, it was Duesenberg in 1921 who pioneered the use of Malcolm Loughead's four-wheel hydraulic brakes. Hydraulic brakes produced quicker, more balanced braking with less effort. However, the Duesenberg system was plagued with a leak problem that gave the brakes a bad name until Chrysler introduced its much improved four-wheel hydraulic brake system in 1924.

Firestone gave road traction a hearty boost when it introduced the low-pressure "balloon" tire in 1922. The conventional tire of the time carried a 345 to 380 kPa inflation pressure, which gave it a small footprint and rock-hard ride, thereby lending itself rather easily to skids. The balloon tire reduced inflation pressures to 240 to 275 kPa. It had a greater footprint and thus a better grip on the road. Within three years, balloon tires became standard on most cars.

Test technology advanced markedly with the opening of the General Motors Proving Ground in 1924 on a 460-ha site near Milford, Michigan. It was the first such large-scale facility set aside by a manufacturer. Before the year ended, engineers invented the first brake test decelerometer, followed in 1926 by the water trough and in 1929 by the Belgian block road.

At the federal level, Herbert Hoover, then Secretary of Commerce, conducted the National Conference on Street and Highway Safety in December 1924. The conference resulted in a number of committees to explore the problems of traffic safety. Two years later, in March 1926, a second conference was held to assess the progress made during the interim. From this latter conference came the Uniform Vehicle Code, the first document to provide guidelines for the safety of road and vehicle.

The 1930s to World War II: Struggling for Survival

While significant advances were made in areas such as body structure, handling, and lighting, most automobile manufacturers had little energy and cash; they merely survived during the Great Depression as people simply stopped buying cars.

In 1935, General Motors introduced the Turret-Top, a feature that Harley Earl, head of GM Styling, claimed to have perfected. The Turret-Top essentially completed the all-steel body by adding the roof to its steel construction. Prior to the Turret-Top, steel bodies terminated at the roof line because the industry had not discovered a way to include the roof in its stamping operations. Closed cars of the era carried roofs made of wood bows covered by fabric. Earl, in designing his Aero-Dynamic Coupe for the 1932 Chicago Exposition, explored the possibility of an upper structure completely made of steel. However, the flat steel roofs that he devised always set up a drumming sound during road travel. He finally cured this problem by crowning the roof surface. GM management was so pleased that it brought the concept into reality and named it the Turret-Top.

In 1934, Fisher Body began engineering the one-piece, all-steel Turret-Top.

Obviously, a steel roof was much safer for auto occupants. Without it, it was not unusual for a passenger to be expelled through the wood/fabric roof in a severe collision.

Ride and handling benefited from the move toward independent front suspension on many models in 1934 and the recirculating ball steering gear in 1940. With independent front suspension, the vehicle no longer lost steering capability if the traditional solid front axle, held in place by stiff springs, had collapsed on it. The recirculating ball steering gear, introduced by the Saginaw Steering Gear Division of General Motors on the 1940 Cadillac, markedly reduced steering effort by adding small steel balls within the grooves of the screw and nut that recirculated back and forth as the driver turned the steering wheel.

Another significant safety contribution of the decade was the sealed beam headlamp that brought order to automotive lighting systems. Prior to the sealed beam, during the early 1930s, as many as 36 different lighting systems were on the market, resulting in enormous confusion over replacement parts and proper aiming instructions. Working together in unparalleled harmony, the automobile and lighting manufacturers combined to produce the sealed beam headlamp in which the reflector was permanently fused to the glass to prevent the intrusion of dust and moisture. Moreover, the sealed beam headlamp produced a 50% increase in candlepower. Initially introduced in 1939, the new headlamp was adopted by virtually every auto manufacturer within a year.

Another industry-adopted breakthrough was the safety rim wheel introduced by Chrysler on its 1941 automobiles. The safety rim, perfected by Owen Skelton, added a simple hump on the inside of the rim that held the tire in place regardless of the amount of air pressure inside the tire. It was designed primarily to keep the tire on the wheel during high-speed blowouts, using centrifugal force to hold the tire firmly in place.

Other advances of the 1930s were the appearance of inside sun visors on several models in 1932, safety padding on front seat backs and recessed instrument panel controls on 1937 Chryslers, windshield washers on the 1937 Studebaker, front and rear directional signals with a self-canceling feature on 1940 Buicks, the double latch to prevent hoods from flying open while driving (on 1941 Chryslers), and the first day/night inside rearview mirror on the 1942 Cadillac.

Crash testing of a crude fashion was explored by General Motors at its Proving Ground in 1934. It consisted of a test driver standing on the running board of a GM car and directing it down a hill into a wall of concrete blocks, jumping aside immediately before it hit. Chrysler also pushed an Airflow off Bald Mountain. Both cases were to evaluate damage, the only kind of analysis that could be made at the time.

Outside the industry, several interesting developments occurred. In 1933, Amos Neyhart gave the first high-school driver education course at State College High School in Pennsylvania, and the Evanston, Illinois, police force opened the first traffic officers training school. In 1937, the Joint Committee on Uniform Traffic Control Devices published the first Manual on Uniform Traffic Devices, which standardized sign appearance and language.

The Automobile: A Century of Progress

The Federal Aid Highway Act of 1938 saw Congress direct the chief of the Bureau of Public Roads to institute a feasibility study regarding the building of six interstate roads across the nation. The results of the study were not acted on until after World War II, but they were no doubt influenced by the success of the Pennsylvania Turnpike, which opened to traffic on October 1, 1940.

The First 100 Years of Transportation Safety: Part II

Anthony J. Yanik

Transportation safety has been defined in various ways over the years, each succeeding generation being faced with the task of improving on the efforts of predecessors. In the early days, transportation safety was considered the domain of the driver. It was assumed that automobile manufacturers had given the operator the appropriate tools for driving safely, such as reliable and durable brakes, steering, lighting, wheels, and suspension systems. It was up to the driver to use these tools in a manner that would safeguard the driver and passengers, as well as occupants in other vehicles, bicyclists, and pedestrians.

Over the past four decades, however, the technology for improving automobile crash protection and crash avoidance has advanced remarkably. In response, the number of vehicle fatalities has decreased more than anyone would have been willing to predict in the 1950s, especially in the face of the growing numbers of vehicles on the road.

Post World War II: The 1940s and 1950s

One of the byproducts of World War II was the development of techniques for studying human tolerance to injury, pioneered by the Air Force. Such study was viewed with interest by the automotive industry and others because of a lack of understanding of the dynamics of, and the means of simulating, accidents and recording them for later study.

In 1949, Corporal Elmer C. Paul of the Indiana State Police developed a technique for investigating the cause of accidents, which soon spread across the state. Paul found himself designated as head of Indiana's Auto Crash Injury Department, the first of its kind.

Paul's work eventually attracted the attention of Cornell Aeronautical Laboratories, which had been studying airplane crashes for several years and hoped to expand its interests to automobile accidents. A working relationship between Cornell and Paul was established which, by the fall of 1950, resulted in the design of the first formal accident report form that provided data on both car damage and occupant injury.

Cornell also began experimenting with the first snubber device to bring an automobile to an abrupt halt in much the same manner as a barrier impact but without causing exterior damage to the car. At the time (the early 1950s), the actions of a dummy during such impact could be recorded by newly developed techniques for high-speed photography, but the technology for instrumenting dummies and measuring impact forces on parts of the dummy's anatomy had not been perfected. Cornell's dummy was a crude figure made of iron pipe and balsa wood. The tests actually were done under contract to the Liberty Mutual Insurance Company looking toward designing a safety demonstration show car.

An improved version of the snubber was engineered by General Motors and installed at the Milford Proving Ground in 1955. The first full-scale crash tests utilizing instrumented dummies to record decelerations and strains were pioneered by the University of California at Los Angeles (UCLA) Institute of Transportation and Traffic Engineering around 1951. Initially, modified department store dummies were used, later to be substituted by more refined specimens, one created by UCLA and another by Sierra Engineering.

In 1954, Ford became the first in the industry to construct a crash test facility, but the lack of good dummy simulators continued to plague anyone working in this area. Also in 1954, General Motors pioneered the inverted head form impact test to evaluate instrument panel tops. In the following year, 1955, Ford introduced the Blak Tufy dummy, a torso-shaped body block suspended on a pendulum and swung into steering wheels at various speeds to evaluate impact forces. The Blak Tufy continues to be used today as a requirement for Safety Standard 203, Impact Protection for the Driver from the Steering Column System.

The product safety innovations that followed World War II were numerous and by the 1950s began to focus more on protective devices. Studebaker led the post-war way with self-adjusting drum brakes on its 1946 models. Self-adjustment meant that an appropriate lining-to-drum clearance was continually maintained as the linings wore.

In 1948, B.F. Goodrich introduced tubeless tires, which ran much cooler than tube tires at high speeds and were more apt to respond to a puncture with a "slowout" rather than a blowout. However, it was not until the 1955 Packard that tubeless tires were offered as standard equipment.

Chrysler made safety news with its 1949 models by adding sponge rubber covered by leather across the top of its instrument panels as a rudimentary form of head protection. Chrysler also introduced the pressure vented radiator cap that year.

Following soon thereafter were two abortive attempts at disc brakes: one by Crosley on its little Hotshot sports car patterned after the aircraft-type disc brake, and the other a self-energizing hydraulic system by Chrysler on its 1949 Imperial. The latter consisted of two flat pressure plates to which were bonded brake linings. Braking action occurred when the pressure plates were forced outward against the brake housings. Both the Crosley and Chrysler systems were unable to dissipate heat effectively and soon disappeared from the market. However, in Britain, Dunlop patented the disc brake as we know it today.

The most effective safety device yet to be seen, the seat belt, was introduced by Nash in September 1949. The seat belt was meant by Nash as a safety measure to accompany its introduction of the reclining front seat.

Nash quit providing the seat belt with the seat after 48,000 were built, at which point the company discovered that only about 1,000 of the belts were being used. However, Nash did feature the belts on its 1951 show cars. It was Cornell Aeronautical Laboratories that kept the issue of seat belts alive, having issued a report in 1953 on the results of its tests on aircraft-type seat belts which put to rest many of the questions and doubts the industry had regarding their worth. By 1954, manufacturers began testing various types of seat belts offered by the supplier industry—too many types, in fact—which brought the Society of Automotive Engineers (SAE) into the picture. SAE formed a committee to develop appropriate belt strength and performance criteria.

The Lifeguard safety package that Ford Motor Company introduced on its 1956 models was given wide publicity. Standard safety features were a depressed steering wheel hub that allowed the rim to distort under impact without the driver contacting the hub, a double ball (for deflection), a shatter-resistant inside rearview mirror, a double grip door lock, and stronger seat belt anchors. Optional were front and rear lap belts that could withstand 17,800 N of force, padded sun visors, and a padded instrument panel top. Unfortunately, Ford cars did not sell as well as General Motors cars that year, giving rise to the canard that "safety doesn't sell."

Virtually all automakers introduced interlocking door latches in 1956, in direct response to the findings of Cornell Aeronautical Laboratories, whose accident data studies listed ejection as the leading cause of automotive fatalities. Other safety innovations for this period included the

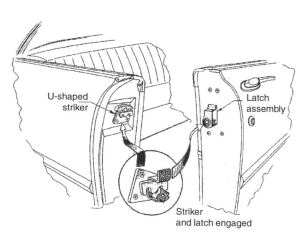

The Lifeguard safety package available on 1956 Ford Motor Company models included this double grip door lock. (Source: SAE 560310.)

U-shaped striker

Latch assembly

Striker and latch engaged

introduction of the ball joint front suspension to prevent front end diving on the 1952 Lincoln, the first attempt at a shoulder belt on the 1957 Chevrolet, and the first rear window defogger and child-guard rear door locks on 1957 Chryslers. (The first, true, three-point lap/shoulder belt became a standard feature on Volvo cars in 1959.) In 1958, Oldsmobile introduced the flanged brake drum for faster brake cooling, and several General Motors models led by Buick added folding front seat-back locks to their two-door models. In 1959, American Motors introduced the head restraint and Chevrolet introduced the four-way hazard warning flasher, both staple safety items today, while Chrysler explored the first automatic dimming day/night rearview mirror.

In the public sector, President Truman approved the initial President's Highway Safety Conference held in 1946. At least 2,000 delegates attended from across the country. They produced a cross-discipline set of safety goals, one being that each state appoint a highway safety agency. In 1954, President Eisenhower held a second conference on highway safety to evaluate efforts forthcoming from the 1946 meeting and to inject new life into those efforts that had not been overwhelmingly received.

The Federal Aid Highway Act of 1956 ushered in our modern, high-speed but safe road system. Congress authorized the expenditure of $25 billion to build 66,000 km (41,000 mi) of interstate roads on a 90/10 federal/state payment ratio. The Act established the Highway Trust Fund to finance the system based on highway use taxes. To date, the interstates are the safest roads on which to drive.

Also in 1956, the House Interstate and Foreign Commerce Committee appointed Ken Roberts of Alabama to chair a special subcommittee dealing with automobile safety. The committee was to meet throughout the next nine years as it probed into such areas as law enforcement, driver training, licensing, and road and vehicle design.

The 1960s and the Highway and Vehicle Safety Acts

Prominent at the outset of the 1960s was the state of Wisconsin, which passed a law requiring that all cars sold in that state, beginning in 1962, must be equipped with seat belts. Before the dust of that legislation had settled, 22 other states had passed similar laws. In response, the industry made lap belts standard equipment on all 1964 models. The Roberts Committee surfaced in the news in 1963 with a bill requiring cars purchased by the federal government in 1965 and beyond to contain safety features such as seat belts, padded sun visors and instrument panels, dual brake systems, outside rearview mirrors, four-way warning flashers, and backup lights.

The middle 1960s became a time of turmoil as Congress, alarmed by the rise in automobile fatality rates, sought to involve the government in the area of vehicle safety. Abraham Ribicoff, chairman of the subcommittee on the Reorganization of Government Operations, used this committee as a forum to raise public opinion for the need of such government intervention, aided by a sympathetic eastern press and the personality of Ralph Nader, one of Ribicoff's aides. When the dust of the hearings had cleared, Congress had passed Public Law 89-563, the National Traffic and Motor Vehicles Safety Act of 1966, thus making the design and manufacture of automobiles a regulated industry. A new safety agency was created within the Department of Commerce, the National Highway Safety Bureau, headed by William Haddon. Within months, Haddon and the new agency had proposed 22 safety standards, 20 of which were issued in final form on January 31, 1967, to apply to all automobiles beginning with model year 1968. To give the Agency more independence, it was transferred on April 1, 1967, to the newly created Department of Transportation, with Alan Boyd as its first secretary.

Overshadowed by the news at the governmental level was the introduction of a number of meaningful innovations regarding occupant protection, especially by General Motors. But Chrysler led the decade by offering the industry's first alternator on the 1960 Valiant. The alternator made it possible to generate electricity although the engine was idling. Within a short time, all manufacturers had shifted from the generator to the alternator for electrical power regeneration.

In 1962, both Cadillac and Studebaker made the dual master cylinder standard equipment so that braking could continue on one two-wheel set of brakes if a malfunction should occur in the other.

The year 1966 was significant for the introduction of high-penetration-resistant windshield glass that would act as a fire net to catch the head if an impact should shatter the windshield. It featured a plastic interlayer twice as thick as its predecessor and a lower bond between the glass and the plastic so that it would have more "give." The new windshield was a product of three years of testing among glass and auto manufacturers. The first cars to receive the new windshield were late-model 1965 Thunderbirds.

In 1967, General Motors pioneered the energy-absorbing steering column that would compress up to 200 mm (8 inches) if struck by the upper torso of the driver during an accident. A section of the column was designed as a wire mesh that would fold on itself similar to a Japanese lantern. This design was subsequently altered by General Motors to one in which one tube telescoped into another. The two tubes were separated by steel balls that rolled grooves into the metal as they telescoped, thus creating an energy absorption action. The EA column appeared on Chrysler, American Motors, and General Motors cars.

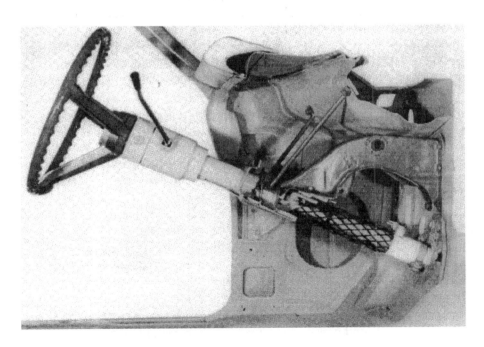

In 1967, General Motors pioneered the energy-absorbing steering column that would compress up to 200 mm (8 inches), folding on itself. (Tenth Stapp Conference.)

In 1969, General Motors introduced the Side-Guard door beam on its full-size models to provide improved protection for occupants from side impact. This innovation soon made its way throughout the industry and became the subject of a safety standard.

Aside from these significant items, the list of new safety features that found their way into vehicles during the 1960s is quite impressive. For example, Ford brought out the manual clutch/electric starter interlock in 1960. In 1962, Cadillac adopted cornering lamps, while most makes adopted amber front turn signals and installed lap belt anchorages for all seating positions. In the same year, Studebaker made front disc brakes standard on its Avanti and optional on all other Studebakers. In 1964, the industry standardized the automatic transmission gear selector pattern, and in 1965 Goodrich began producing the first radial-ply tires for the domestic market. In model year 1966, Chrysler introduced recessed inside door handles. In model year 1967, Volvo added a deformable, recessed, steering-wheel hub, and Ford featured a door latch that would lock automatically at speeds above 13 km/h (8 mi/h). A flurry of activity in 1968 in response to federal standards produced energy-absorbing seat-back tops and side marker lights on all models. In 1969, the contoured windshield header for improved head impact kinematics appeared on many General Motors makes, and automatic locking seat belt retractors were released by Chrysler and General Motors. Ford completed the picture with a rear wheel anti-lock brake option.

Safety test technology saw several advances through the 1960s as well. In 1962, General Motors designed the industry's first high-speed impact sled, capable of simulating actual collision decelerations, and installed it at its Milford Proving Ground. In 1966, engineers at GM Research originated the methodology for determining the extent of injury hazard produced when measuring the forces of impact on instrumented dummies during impact tests. It was from such investigations that the 1,000 HIC (Head Injury Criteria) value was established beyond which significant head injury was considered likely. Finally, in 1969, General Motors finished the first completely enclosed barrier impact test site to permit testing under all weather conditions.

The 1970s and the Air Bag Controversy

Looking back to the 1970s, it seems that the decade consisted of a protracted difference of opinion between government and industry on how, when, or whether air bags should be introduced to the public.

On March 19, 1970, Secretary of Transportation John Volpe announced that the National Highway Safety Bureau was being removed from the Federal Highway Administration and set up as a separate entity with its administrator reporting directly to him. It also received a new name, the National Highway Traffic Safety Administration (NHTSA).

Volpe was so intrigued by the possibilities of air bags as a protective device that he had the NHTSA issue a proposal for the feasibility of requiring air bags on all cars by January 1, 1972. The controversy erupting from the proposal made it quite clear that much concern existed within the industry regarding their technical readiness, their costs to and acceptance by the public, their reliability, and even the unintended side effects of rapid bag deployment. Soon the target date was reset for January 1, then July 1, and even August 15, 1973, before being extended to August 1975.

During the interim, the NHTSA decreed that beginning in October 1973, all cars must have a seat belt starter-interlock system, which would prevent drivers from starting their automobiles if they had not connected their seat belts. It also called for a new type of seat belt system, the combined lap/shoulder belt, which remains common today. Automobile manufacturers found themselves in the unenviable position of engineering complex new belt systems while simultaneously working on defining air bags. Resentment over the enforced usage of belts rose quickly within the public and came to a head in December 1974 when Congress decreed that the starter-interlock be eliminated.

While the interlock controversy was taking place, both Ford and General Motors were diligently at work, attempting to engineer a satisfactory air bag system. Ford took first honors when it installed right front passenger air bags in a fleet of 800 Mercury sedans in 1972. General Motors followed with a fleet of 1,000 full-size 1973 Chevrolets equipped with driver/right front air bags. Results of the fleet test were substantial enough to General Motors to result in a schedule to build 300,000 full-size air-bag-equipped cars over a three-model-year period, 1974 to 1976. Sales were quite poor (no more than 10,000 per year) for various reasons: cost (the initial price was approximately $325), fear both within the public and at the service garage regarding inadvertent deployment, and the fact that the oil embargo of 1974 hit the sales of full-size cars quite hard.

In the spring of 1975, the NHTSA held hearings on the feasibility of the air bag ruling and by 1976 decided to co-fund with industry a demonstration program of 500,000 air-bag-equipped cars over model years 1979 and 1980; however, this decision was abruptly terminated in March 1977 when the new Secretary of Transportation, Brock Adams, held another round of hearings, after which he announced that air bags would be required to be built into all new cars during a three-year period commencing in model year 1982. This was how matters stood as the 1970s came to a close.

On a different front, the Department of Transportation also had become enamored with the thought that, with appropriate encouragement, a vehicle could be designed that would provide crash protection up to 50 mi/h (80 km/h) at no appreciable change in consumer price. The encouragement came in the form of nearly $8 million in federal funds to contractors whose proposed designs were accepted by the NHTSA. On June 26, 1970, the Department of Transportation announced its decision: Fairchild Corporation would be awarded $4,547,500 to complete its proposal, and American Machine and Foundry (AMF) $3,440,000. A third acceptance was that of a bid from General Motors for $1 to produce an Experimental Safety Vehicle (ESV) to government performance specifications. The contract terms for the three producers were that

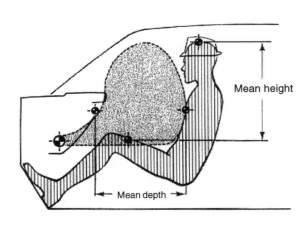

Mean height

Mean depth

Ford installed right front passenger air bags in a fleet of 800 Mercury sedans in 1972.

the ESV would be a five-passenger, 4,000-lb (1,814-kg) car on a wheelbase not to exceed 124 inches (315 cm) with bumpers capable of sustaining impact speeds up to 10 mi/h (16 km/h). The AMF and Fairchild designs were to be delivered in 18 months and the General Motors version in 28 months.

What emerged were three vehicles, the likes of which had never been seen. The AMF concept carried massive front and rear bumpers that gave it the appearance of a double-ended battering ram. Its front pillars were covered in 50 to 75 mm (2 to 3 inches) of foam with more thick foam on the inside door trim. There were air bags all around, and the rear window of the vehicle was designed to eject automatically during impact to relieve the air pressure from all air bags deploying simultaneously. The vehicle also included a roof-mounted periscope for rear viewing.

The Fairchild vehicle featured a roll-cage-type structure designed to pro-vide crash protection from accidents of any direction. The roof pillars, in effect, acted as roll bars. It, too, featured air bags for all seats and had a unique bumper that automatically extended 300 mm (12 inches) forward from the vehicle at speeds over 40 km/h (25 mi/h) to help cushion forward impacts.

The General Motors ESV of the early 1970s had a special frame expected to absorb the force involved in a collision, high-strength center roof supports for rollover protection, and fixed side glass to guard against ejection.

The Automobile: A Century of Progress

The General Motors entry was much more conventional in appearance. It had a special frame expected to absorb the force involved in a collision, high-strength center roof supports for rollover protection, and fixed side glass to guard against ejection. Special thick interior bolsters were introduced to give impact protection at speeds up to 48 km/h (30 mi/h), after which air bags would deploy. Although not under contract, Ford also produced an ESV but to its own performance terms. Also chiming in were Volkswagen, Nissan, Toyota, Volvo, and Mercedes-Benz with ESVs of their design but at half the mass required by the Department of Transportation contract.

Although the Department of Transportation ESVs were interesting, they primarily indicated that, at that time, it would take massive structures to provide the type of impact performance desired, and such structures would drastically affect fuel economy at a time when energy crises and fuel economy standards were becoming a way of life. Whether they could have been built at an affordable price and whether the buying public would have accepted such oddities in the name of safety are questions that will remain unanswered.

One of the most significant governmental acts of the 1970s took place in the state of Victoria, Australia. On January 1, 1971, Victoria passed a law that mandated the use of seat belts. Little mention of it was made in the world press until the first year's results were in, indicating a 21% drop in fatalities. Soon, other Australian states followed Victoria's lead, followed by the European countries of France, Finland, and West Germany, and then the remainder of Europe. In the United States, however, it was felt that the public would never accept such a restriction.

Another significant governmental action occurred on January 1, 1974, when Congress made the temporary 55 mi/h (88.5 km/h) highway speed limit permanent, to take effect in one year. Although the 55 mi/h (88.5 km/h) limit was meant to conserve energy, it also lowered highway speeds to a point where accidents might be more survivable within current technology.

In 1977, Tennessee became the first state to adopt a child restraint use law. Results of the law were watched closely by the balance of the nation and seemed promising enough so that other states began to follow the Tennessee lead.

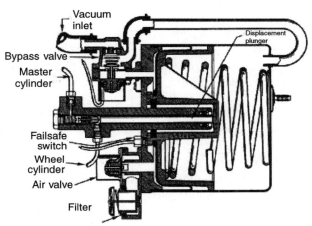

On the product front, several safety innovations that have become commonplace today were unveiled. In 1970, General Motors introduced the infant carrier to protect babies up to one year old being carried within the vehicle, while Ford made news with the first steel-belted, radial-ply tires on the Mark III and Thunderbird. Ford also offered power front disc brakes on many of its 1970 models.

The pressure modulator of Chrysler's Sure Brake System of 1971, the first four-wheel, anti-lock brake system option. (Source: SAE 710248.)

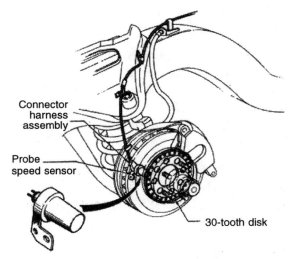

Pontiac introduced the industry to the maintenance-free, sealed battery on its 1971 models. No longer were drivers forced to keep a wary eye on whether or not they had kept their battery reservoirs filled.

The first traction control system, the Max-Trac, became available as an option on the 1971 Buick Riviera. Shown is the front-wheel-speed sensor installation. (Source: SAE 710612.)

Chrysler offered the first four-wheel, anti-lock brake system option that year, the Sure Brake System, and Buick added the first traction control, the Max-Trac, which became available as an option on the 1971 Riviera.

In model year 1972 Oldsmobile introduced the first disc brake audible wear indicator on the Toronado, and all vehicles came equipped with seat belt buzzers and warning lights that signaled if the vehicle occupants had not put on their seat belts.

Per federal mandate, all passenger cars were required to carry 5 mi/h (8 km/h) bumpers, front and rear, for 1973. The rule ostensibly was meant to protect surrounding safety equipment from low-speed impacts but was generally thought to be a low-speed damage cost control device. Also in 1973, all cars were required to have fire retardant interior materials per federal safety standards. In the same year, General Motors full-size models were equipped with the first front suspension ball joint wear indicator, allowing a visual check to be made of the need for ball-joint replacement.

For model year 1976, Volkswagen introduced the automatic shoulder belt cum knee bolster to the U.S. scene, followed by Chevette in model year 1978. However, the following year, Chevette brought out the first continuous loop, automatic lap/shoulder belt as an option.

In 1978 the Safety Administration changed its requirements for headlamp performance, raising permissible high-beam light output from 75,000 to 150,000 candlepower. The change permitted U.S. manufacturers to begin equipping their 1979 automobiles with halogen sealed-beam headlamps, which were common in Europe but not previously permitted in the United States. First to offer them was the 1979 Lincoln Versailles.

Also in 1978, General Motors announced that it had developed a new test dummy, the Hybrid III, which was markedly superior to the Hybrid II dummy (which General Motors also had designed to test the 1974 air bag systems).

The 1980s: Passive Restraints Arrive

The on-again/off-again air bag legislative process continued as President Reagan took office. Late in October 1981, the NHTSA rescinded the rule promulgated during the Carter administration that all cars be equipped with passive restraints beginning with full-size cars in 1983. However, in August 1982, the U.S. District Court of Appeals overturned the NHTSA rescission of the rule and ordered the auto industry to begin installing passive restraints in model year 1984, pending a review by the Supreme Court in June 1983. The Supreme Court in turn upheld the decision of the lower court.

On July 11, 1984, Transportation Secretary Elizabeth Dole issued new rules that would require the phasing in of automatic crash protection beginning with 10% in model year 1987, 25% in 1988, 40% in 1989, and 100% in 1990. Realizing that the auto industry was more likely to opt for the economical automatic seat belt solution for meeting the mandate, Dole dangled the carrot of 1.5 credits for each car equipped with air bags and added further that she would consider eliminating the requirement if states representing two-thirds of the population would adopt seat belt use laws by April 1, 1989.

On July 12, 1984, New York became the first state in the nation to adopt a seat belt use law, followed in January 1985 by New Jersey and Illinois. By the end of 1985, 17 states (encompassing 54% of the population) had passed seat belt use laws.

Meanwhile, Mercedes-Benz announced that it had sold 17,000 1985 cars equipped with a driver-side air bag and would make it standard on all of its 1986 models. Ford also had offered a special safety package that included a driver air bag on its 1985 Tempo and Topaz, and sold 7,400 of them. Ford then announced that it would make driver air bags an $815 option on all Tempo/Topaz four-door sedans beginning in the spring of 1986. Supplier difficulties limited the option to 17,000 vehicles, but all were sold. Ford then said it planned to offer driver air bags on more than 750,000 models in the 1990 model year, including the Taurus/Sable. Chrysler also announced that it would have driver air bags in more than half its models (600,000) that year.

However, as the initial 10% part of the passive restraint rule began to take effect in model year 1987, the industry did indeed meet it, primarily with automatic lap/shoulder belts. In addition, General Motors said that it would begin to replace outboard rear seat lap belts with lap/shoulder belts over a three-year period, a practice that in time extended to other manufacturers. In 1989, Lincoln Continental became the first to equip its models with driver and front passenger air bags.

Digressing momentarily from its activity on passive restraints, the Safety Administration published a new rule requiring center high-mounted stop lamps on all cars built after September 11, 1985. In 1987, it instituted the New Car Assessment Program (NCAP) in response to Title II of the Motor Vehicle and Cost Savings Act of 1972. The NCAP tests consisted of head-on crashes into a fixed barrier at 35 mi/h (56 km/h) with measurements taken of impacts to driver and front-seat-passenger dummies' heads, chests, and upper legs and compared to predetermined thresholds for potential real-life injury. The NCAP tests have been a matter of controversy to date as to whether they represent the results of real-world crash situations.

Safety products introduced during the 1980s were the first pre-crash belt tensioner by Mercedes-Benz in model year 1981, the door-open warning light by Volvo in 1982, the automatic day/night inside rearview mirror by Ford in 1983, the front/side window defogger in the Chevrolet Spectrum in 1985, and the introduction of advanced electronic four-wheel, anti-lock brakes (standard on BMW and optional on the Lincoln Continental and Mark VII) also in 1985. Shoulder belt height adjusters first appeared on the Mercedes 300E and the Audi 5000CS Turbo in 1987, and the automatic transmission shift lock by Audi was introduced that year.

The 1990s: Looking Forward to the Millennium

Apparently satisfied with the progress it had made in passive restraints to protect against frontal crashes, the Safety Administration now turned its attention to side impacts. In October 1990, it brought out a new rule calling for dynamic side-impact tests in which a moving barrier would impact a test car. The then-current side-impact test was a static test in which a steel cylinder was forced slowly against the side of a vehicle on a test bed.

This calling of attention to side-impact protection ultimately created interest in the development of a side-impact air bag system. Volvo became the first to introduce such a system on its 1995 850 series. Others are hard at work to develop side air bag designs of their own, some issuing from the seat and others from the head restraint or the door.

Also in the 1990s, Chrysler became the first to introduce the integrated child seat, offering it as an option in its 1992 minivans. To increase the visibility of oncoming vehicles in daytime traffic, General Motors in 1995 began installing Daytime Running Lamps on several Chevrolets and Chevrolet/GMC pickup trucks.

And the beat goes on. There is no doubt that new safety developments will occur throughout the remainder of the 1990s. It is only a matter of time before devices such as front and rear obstacle detection or collision avoidance systems, high-intensity-discharge headlamps, cruise controls that keep an automobile a set distance behind another vehicle, or night vision systems become a commercial reality.

Meanwhile, the auto industry can look back with pride at what it has accomplished during the past 100 years, a century in which the driving experience changed from an adventure in which safety was left to the skills of the driver to the present-day drive, in which those skills are coupled with new product innovations that not only decrease the possibility of a collision but also decrease the severity of injury should one inadvertently occur.

A History of Automobile Electrical Systems

Ralph H. Johnston
Delphi Energy and Engine Management Systems (Retired)

The electrical system is the "nervous system" of the automobile and is essential to the car's "pleaseability" to its owner. It is essentially transparent to the driver unless a problem develops, at which time he/she becomes painfully aware of its importance. The evolution of the electrical system is divided into several time periods, and the important developments of each major part are covered for that particular time period.

Up to 1912

Ignition

As the internal-combustion reciprocating engine developed in the late 1800s, one of the major challenges was initiating air/fuel mixture combustion in the combustion chamber. Many ignition methods were attempted, including open-flame, hot-tube, and hot-wire igniters. None of these efforts was entirely satisfactory because the timing of ignition was highly variable and they were difficult to implement. Creating an electric spark

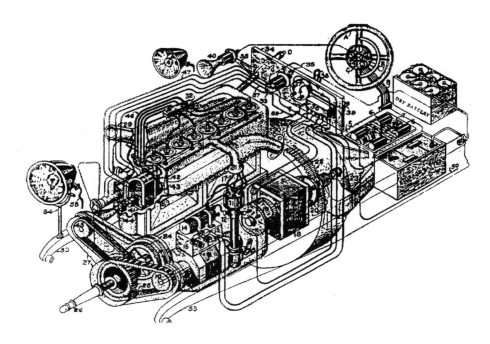

Electrical system of a modern automobile.

within the combustion chamber proved to be the most satisfactory method to ignite the mixture, and thus began the effort to develop suitable implementation.

An early mechanism to produce the spark in the combustion chamber consisted of a metallic arm in the chamber attached to a shaft which extended through the chamber wall from outside the engine. The arm made intermittent contact with a stationary rod installed through the chamber wall in an insulating bushing. A cam synchronized with the crankshaft drove a bell crank that rotated the contact arm in the chamber through a small angle, making and breaking the contact between the arm and the rod. A low voltage was applied, from a series of dry cells, to the contact rod through an inductor. When the contact in the chamber closed, a current would build up in the inductor. When the contact was opened, a spark discharge occurred, igniting the mixture. This low-voltage ignition system was used on multicylinder and single-cylinder engines. However, the terrible environment for electrical contacts in the combustion chamber caused their rapid deterioration, and continuous maintenance was required. Also, control of ignition timing was difficult.

The next step in the search for improved ignition was to place a spark plug in the combustion chamber, with a gap across which the high energy could be discharged. This, however, required very high voltage to ionize the spark gap under the high combustion chamber pressure existing at the time of ignition. It was accomplished by the development of a transformer with a very large ratio of secondary to primary windings. The voltage of the primary winding was 6 or 12 volts, and the voltage of the secondary was greater than 20,000 volts, which was adequate to ionize the mixture in the spark plug gap.

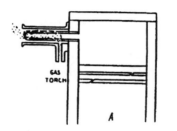
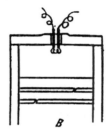

(A) Hot-tube igniter. (B) Hot-wire igniter.

Because a transformer depends on alternating current, a means had to be provided to produce the alternating or pulsating current in the primary winding. This was initially accomplished by an alternating current generator called a magneto. There were two types of magnetos: a low- and a high-tension (voltage) unit. The high-tension magneto had a transformer wound on its armature, and its output could be used to fire

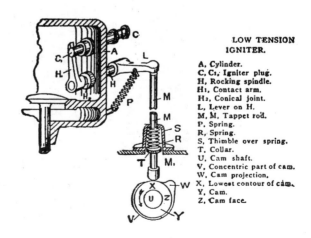

LOW TENSION IGNITER.

A, Cylinder.
C, C₁, Igniter plug.
H, Rocking spindle.
H₁, Contact arm.
H₂, Conical joint.
L, Lever on H.
M, M, Tappet rod.
P, Spring.
R, Spring.
S, Thimble over spring.
T, Collar.
U, Cam shaft.
V, Concentric part of cam.
W, Cam projection.
X, Lowest contour of cam.
Y, Cam.
Z, Cam face.

In-cylinder make/break igniter.

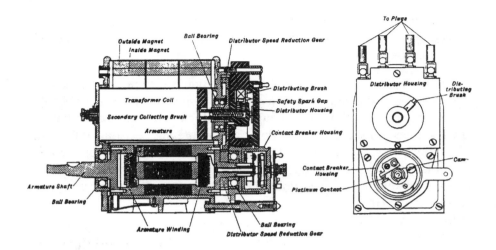

Typical high-tension magneto.

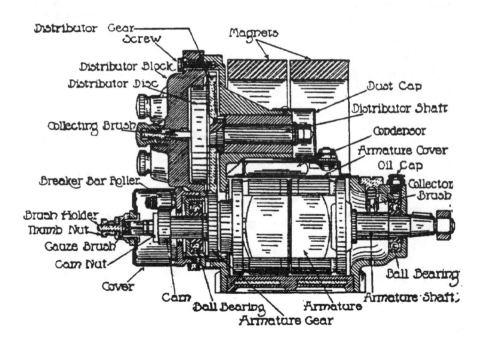

Typical low-tension magneto.

154

the spark plug without further conditioning. The low-tension magneto produced a low voltage which was supplied to the primary winding of a separate transformer, resulting in a high-voltage output from its secondary winding sufficient to fire the spark plug. The output voltage from the magneto was too low during cranking to reliably fire the spark plug; therefore, it was necessary to power the system from a battery, usually a series of dry cells, while hand cranking the engine. Both systems were produced in volumes needed for the early cars.

Another system used extensively in the early days was a "vibrator" or "trembler" to chop the current in the transformer primary to produce the pulsating current. This consisted of a set of vibrating contacts responding to the magnetic field of the ignition coil core to make and break the primary circuit. Other systems used a vibrating contact coil for each cylinder and switched the power supply voltage appropriately to the primary of each coil with a timer driven from the camshaft.

Ford used the coil-per-cylinder system until 1928. Ford also used a unique type of magneto built into the flywheel. It consisted of vee-shaped horseshoe magnets, with the open end of the vee facing outward. The windings of the stator were mounted in a circle at the periphery of the flywheel and were attached to the engine block.

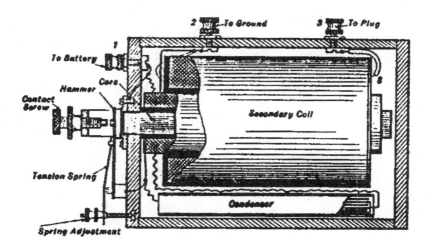

Vibrating, "trembler" ignition coil.

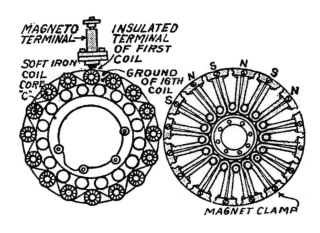

Ford flywheel magneto.

High-tension magnetos generated high voltage using a transformer wound on the rotor of the machine in some cases. In other cases, the rotor had a low-voltage winding which supplied power to a transformer integrated into the magneto. The high voltage fed the center contact on a high-voltage distributor, also integrated into the magneto, which in turn supplied each spark plug with spark energy in the proper firing order. The magneto shaft was driven synchronously with the engine.

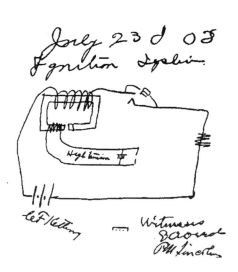

Copy of the original sketch of the breaker-type ignition system.

Some of the suppliers of both low- and high-tension magnetos of the time were Bosch, Connecticut Telegraph and Electric Co., Remy Electric Co., and Splitdorf. Ford supplied its own systems.

A major development in ignition systems occurred in 1908, when Charles F. Kettering developed the single-spark, breaker-type ignition system. It was developed to overcome the deficiencies of the magneto and vibrating coil systems, and was eventually labeled the "Kettering system." Instead of a vibrating contact chopping the current to the primary circuit of the ignition coil, it used a cam-driven set of contacts to interrupt the primary current for the spark event. A single spark occurred as the contacts opened to ignite the air/fuel mixture, instead of the continuous stream of sparks that the vibrator coil produced. Because the ignition system was powered by either dry cells or a small storage battery, power conservation was an important design consideration for the system. This resulted in two variations of the breaker system: "closed circuit" and "open circuit."

The closed circuit configuration kept the primary circuit energized continuously until the spark event, at which time the current was interrupted. The open circuit (make-and-break) types energized the coil primary circuit immediately before the spark event was to occur, allowing enough time for the required energy to be stored in the magnetic field of the ignition coil to produce an adequate spark. This system required less average power than closed-circuit systems. Numerous variations of these breaker-type systems were produced by several suppliers. The implementation of the breaker systems generally combined the functions of the timer, or circuit-breaking mechanism, and the high-voltage distributor in the same assembly. The coil was usually mounted separately in the engine compartment. However, Westinghouse Electric Company produced an integrated unit with the coil in the distributor assembly. Capacitors were required to shunt across the breaker contact points to absorb the energy associated with breaking the primary current. This reduced arcing with its potential for damaging the contact points. Capacitors were sometimes integrated into the distributor assembly; other times, they were external to it.

Several methods were used to drive the distributor synchronously with the engine crankshaft. The speed ratio between the two was determined by the number of lobes on the distributor cam in relation to the number of cylinders of the engine. Eventually, it became common to take advantage of the 2:1 ratio between the camshaft and the crankshaft, and use a 1:1 ratio between the camshaft and the distributor shaft. The number of lobes on the distributor cam was then equal to the number of cylinders.

Batteries

The batteries to power the ignition system were either the dry-cell or lead-acid type. The dry cells, of course, had to be discarded when spent, but the storage batteries could be recharged. This generally required removing the battery from the car and taking it to a charging facility.

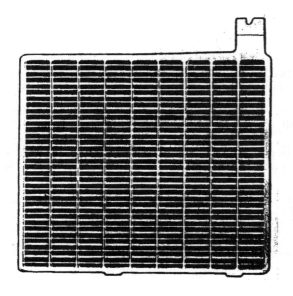

Typical automotive battery plate.

The earliest lead-acid battery "accumulators" were invented by Gaston Plante in 1859. They were made by rolling thin strips of lead foil, with porous insulating material between them, into a "jelly roll" shape. This cylindrical cell was submersed in dilute sulfuric acid electrolyte. Many charge/discharge cycles were necessary to form the batteries to the point at which significant energy could be stored.

In 1881, Camille Faure invented a version of the lead-acid battery which substituted a flat lead grid structure in place of the lead foil. These grids were cast as a flat lattice, into which a lead oxide paste was pressed, forming a plate. The plates were stacked to obtain the desired performance. This construction significantly shortened the time of formation, making the battery more producible in larger quantities. Automotive batteries up to the present time used this basic type of construction, although numerous refinements have occurred in the intervening years.

An energy capacity of 60 A•h was common for powering only ignition. When incandescent lamps came into use, replacing oil and

Strap →

Typical battery-plate stack.

158

acetylene lamps, the battery capacity was typically increased to approximately 120 A•h. The batteries delivered comparatively low power, however.

The component parts of the typical lead-acid storage battery included the lead-antimony alloy plate grids (antimony was used to stiffen the cast lead grid so it could be handled more readily), into which a lead oxide paste was pressed. Each battery manufacturer had its own secret formulations for the ingredients that made up its paste. Plates were stacked to form cells with connection lugs placed at opposite upper corners of the stack. Insulating separators were installed between the plates as they were stacked. The insulators were typically thin layers of wood or porous rubber. The rows of lugs at the opposite corners of the stacks were melted (burned) together into a heavy lead strap to connect all the positive plates to one strap and the negative plates to the other. Generally a lead post was burned to each strap to bring the connections outside the cell where they were accessible. The stack of positive and negative plates (a cell) were placed in a hard rubber jar and submersed in a dilute solution of sulfuric acid. A hard rubber cover

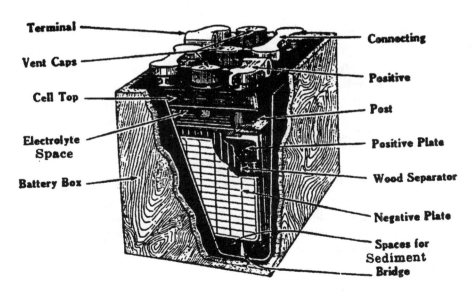

Wooden-case battery.

was sealed to the top of the jar, with the cell terminal posts extending through and sealed to the cover. An access hole was placed in the cover to vent the cell and to permit the addition of water to maintain the specific gravity of the electrolyte.

Each of these cells produce a voltage of approximately 2.25 volts. To make a battery, several cells were assembled into a wooden box, the number of cells dependent on the desired battery voltage: three cells for 6 volts, six cells for 12 volts, and twelve cells for 24 volts.

Until this time, the electrical loads consisted primarily of the ignition system and the lights. No generator was yet available on cars; therefore, the storage batteries had to be recharged periodically, and the dry cells, if used, had to be replaced as needed. However, a new era in the history of the electrical system was about to begin, with the invention of the electric starter.

1912 to 1930

Starters

A watershed year in automotive electrical systems, and in the industry, was 1912, when Cadillac Motor Car Co. introduced the electric self-starter. This had a major impact on the convenience and acceptability of the automobile to the general public. Until this time, the principal means of cranking an engine was the hand crank. The crank was permanently installed in the frame of the vehicle in a bearing, in such a way that it could be pushed into engagement with the crankshaft when cranking the engine was necessary. Cranking an engine took considerable effort, particularly when the engine was cold; thus, few ladies would attempt the task.

Many other methods of starting without the hand crank were attempted because hand cranking could be dangerous. These included igniting the residual air/fuel mixture in the combustion chamber after the engine had stopped, priming the cylinders with gasoline and igniting it, and injecting acetylene gas into the cylinders and igniting it. Another system injected compressed air into the cylinders timed in such a way that it would rotate the engine.

An elaborate system was developed to inject an appropriate mixture of air and fuel into the combustion chamber under pressure. The injection was timed to match the engine firing order, making the engine able to run on this starting system until switched to the conventional carburetor and ignition system.

It is apparent from the level of effort applied to various means of engine cranking that replacing the hand crank was a high priority with developers of the day. As mentioned, hand cranking could be dangerous because the spark timing was manually controlled by the operator, generally with a lever located below the steering wheel on the steering column. If the operator neglected to retard the spark so that ignition occurred after top-dead center, the engine could fire before top-dead center, producing a torque pulse in the opposite direction to that of the cranking effort. This would at minimum jerk the crank out of the operator's hand and could cause a broken arm or more serious injury.

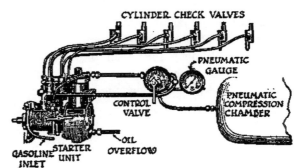

Early gasoline/air starter.

A cranking injury that ultimately proved fatal to a friend of Mr. Henry Leland, head of the Cadillac Motor Car Co., prompted Leland to approach Charles F. Kettering of the Dayton Electrical Laboratory Co. (DELCO) about attempting to develop an electric self-starter for cars. Mr. Kettering's company had been supplying ignition systems to Cadillac, and considerable mutual respect had developed between the two men. The development effort by Kettering was carried out during 1910 and 1911, and was based on work he had done for the National Cash Register Co. several years prior to that time.

The unique requirement for both applications was that the operation of the motor was highly intermittent; therefore, more power could be delivered from a small motor than was possible in a continuous-duty application.

A series configuration of the field and armature windings was selected because of the high stall torque capability of this configuration and its wide operating speed range. The development was completed in 1911, and the system was introduced by Cadillac in the 1912 model. This was a highly integrated unit in which the machine had both starting and generating sets of windings on one armature, each with its own commutator and brushes. Thus, the same machine started the engine, generated electric power to supply the lighting and ignition systems, and kept the battery charged.

The ignition system was also included in the assembly. When cranking, the motor was engaged to a ring gear on the periphery of the flywheel through a speed-reducing geartrain. The shifting of the starter gear into mesh with the flywheel gear was accomplished by a mechanical linkage moved with a foot pedal by an operator. The same mechanism lowered a brush onto the starting winding commutator in the motor to apply the power. This eliminated the need for a high-current mechanical switch for the cranking function. When in the generating mode of operation, the armature was driven from an extension of the water-pump shaft, because

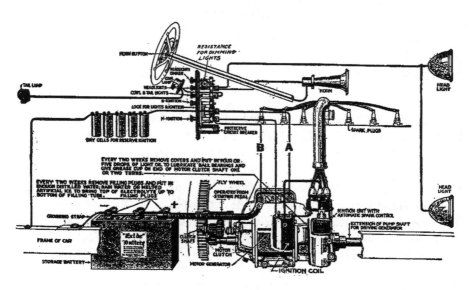

Cadillac 1912 electrical system.

this shaft was driven directly from the crankshaft, and synchronized with it. The distributor for the ignition system was incorporated into the same assembly.

So great was the impact of the electric self-starter on the automotive industry that, within the next two years, 90% of all cars built in the United States had one. Several of the systems used motor/generators of various types. Some combined both starting and generating windings on the same shaft, such as the DELCO system; others placed two separate machines in the same housing. The use of motor/generators continued until 1923, although the use of separate machines for starting and generating began prior to that time.

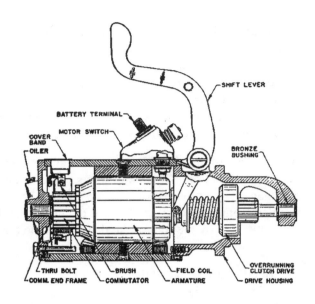

Starter with mechanical engagement.

In 1917, there were approximately 20 suppliers of starting and generating devices, some of which supplied complete electrical systems. Westinghouse Electric Co., DELCO, and Autolite accounted for approximately 40% of all applications. Adding Remy Electric

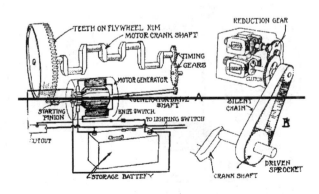

(A) Typical motor generator configuration.
(B) Two separate machines in one housing.

Co., Bijur, and Gray & Davis constituted another 30%. Ford continued to produce its own electrical systems which, of course, represented significant volume in addition to the other suppliers. Remy Electric Co. was purchased by United Motors Corporation in 1916 and through it became a part of General Motors in 1918. In 1926, DELCO was merged with the Remy Electric Co. to form the Delco Remy Corporation, which eventually became an operating division of General Motors.

As time passed, the development of the starter became independent of the generating function. One major focus of the design effort was on methods of engagement of the motor with the flywheel ring gear. Two methods resulted from the work: one moved the pinion gear into engagement with a lever operated by the foot of the driver, and the other was the Bendix Drive.

The Bendix Drive placed the pinion gear on a spiral thread attached to the motor shaft. As the motor began to turn, the inertia of the pinion tended to hold it from turning and caused it to move along the thread until it engaged the flywheel ring gear. A large coil spring cushioned the shock of the engagement process. As the engine began to run, its speed was so much

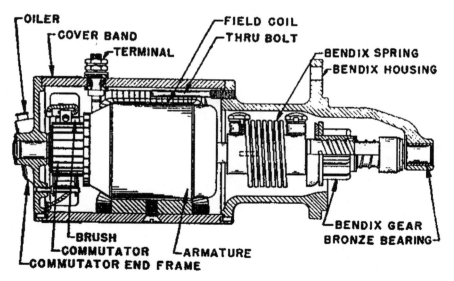

Starter with Bendix Drive mechanism.

higher than the cranking speed that it would spin the pinion back along the thread in the reverse direction, disengaging it from the ring gear. These two methods of engaging the starter to the engine ring gear were most common during this period.

Generators

The development of the generator was coincidental with the electric starter. As mentioned, some of the first generators used the same machine for both functions. Some had both the generating and starting windings on the same armature; others designed two separate machines into the same housing. In the case of the motor/generator on the same shaft, a rather complex mechanism was needed to couple the armature shaft to the crankshaft. When cranking, a large 20:1 gear reduction ratio was needed to achieve adequate torque. When generating, however, the armature had to be driven at approximately triple the crankshaft speed to achieve the needed output power. Current output of the early generators was between 15 and 20 A. Ingenious methods involving gears and overrunning clutches were developed to couple the machine to the engine.

Because the generator was driven at approximately triple engine speed, it covered the same ratio of minimum to maximum speed as that of the engine. This created problems in two areas. First, at least a 3:1 ratio was needed to achieve adequate power at low engine speeds. An even higher

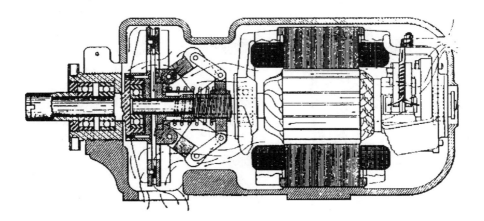

Generator regulated with a centrifugal clutch.

ratio would have been preferred, but a larger drive ratio would turn the machine so fast at high engine speeds that the centrifugal force on the windings would destroy the armature. At best, the drive ratio was a compromise that provided a small amount of power at idle, without causing damage at high speed.

The second issue, caused by the large engine speed range, was the need to provide a means to regulate the output voltage of the generator to match the charging voltage needed by the battery. Without some form of limiting, the voltage and current of the generator increase linearly with speed, assuming the load is constant. Thus, it was necessary to limit the current, the voltage, or both to protect the battery and the electrical system components.

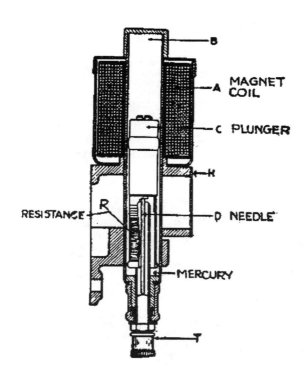

Mercury pool regulator.

The regulation of dc generator voltage provided a fertile field in which to work for developers of the time. Early methods included various forms of bucking field windings to reduce effective field flux in response to increased output current. One method used a centrifugal clutch in the generator drive, which slipped increasingly with speed to limit the speed of the generator and thus its output power. Another regulation method used a centrifugal governor to operate a contact arm on a variable resistor, which increased the resistance in series with the field winding as the speed increased and limited the generator output.

One interesting concept developed by Kettering used a linear solenoid in series with the generator output such that the solenoid plunger moved up and down in response to current. Attached to the plunger was a coil of resistance wire which was in series with the generator field winding. The coil was partially submersed in a pool of mercury; as the output current increased, the solenoid plunger withdrew more of the resistance wire from the mercury. This increased the resistance because less wire was being shorted by the mercury, reducing field current and limiting generator output.

By the late 1920s, most of these early regulators were replaced by regulation using a third brush. To describe the operation of the third brush method of regulation would take more space than is available here; however, briefly, a third brush was placed between the main positive and negative brushes contacting the commutator. The current for the field was supplied by this brush. As generator speed increased, field current would tend to increase, but the position of the third brush in relation to the magnetic poles of the machine caused the flux to be distorted from its normal path, limiting generator output. This method of regulation became common with most suppliers.

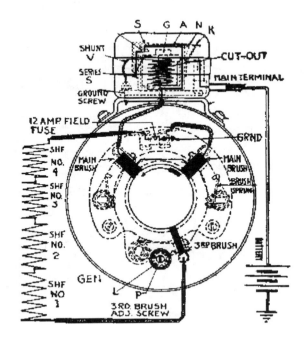

Third brush generator with cutout.

In addition to the third brush method of limiting output current, voltage- and current-sensing relays were developed to improve the accuracy of regulation. The current-sensing relay had its coil in series with vehicle-system electrical load such that all generator output current flowed through it. As the current increased beyond the set point of the regulator, the relay contacts would open, inserting a resistor in series with the field coil and lowering the field current. The reduction in field current allowed the relay contacts to close again, shorting the field resistance and thus increasing the output current. This resulted in the contacts opening and closing at a very rapid rate; thus, the average current was limited to the point at which the current regulator was set.

In addition to current regulation, a voltage-sensing relay was sometimes used for accurate voltage regulation. Its behavior was similar to the current relay, except it responded to the generator output voltage instead of the current. These two types of regulators came to be known as "vibrating contact regulators" because of their rapid opening and closing response to current and voltage. They were both used in conjunction with third brush, current-limiting, generators.

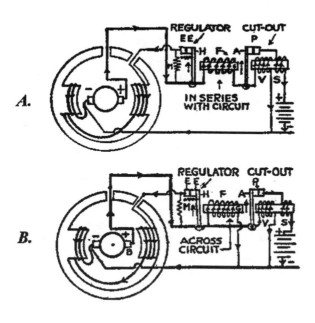

(A) Current regulator. (B) Voltage regulator.

Temperature compensation of the output voltage from the generator was used in very early generating systems. Because of the difficulty in charging a cold battery, it was desirable to raise the charging voltage as the temperature became colder. The early systems used a bimetallic element to operate a set of contacts which inserted or removed a resistor from the field circuit as the temperature changed. Temperature compensation in future systems would become more sophisticated.

One other device necessary for use with the dc generator was the "cutout," a relay in which contacts were in series with the generator output. When the generator was not operating, the contacts were open so the battery could not be discharged through the generator windings. The relay had two windings making up its coil: one was across the output terminals of the generator to respond to its output voltage, and the other was in series with the generator output to respond to its output current. When the car was started and the generator output voltage began to rise, the current in the voltage coil of the relay would increase to the point at which the contacts would close, connecting the generator output to the vehicle electrical system. When the contacts closed, current began to flow to the loads through the current winding of the relay. The flux of the relay reinforced that of the voltage coil, holding the contacts closed even if the voltage was reduced at low generator speeds.

Ignition

By the 1920s, most of the basic invention relative to battery ignition had been done. The Kettering system had become well established and would be predominant in the industry for many years. Most systems consisted of a distributor assembly, which included the cam to open and close the breaker contact points,

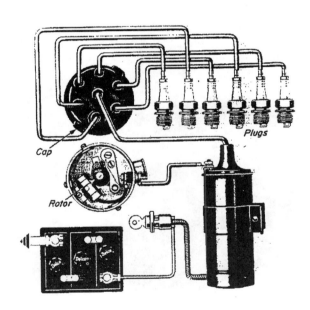

Typical distributor and ignition coil.

the contact points, the high-voltage distribution switch, and a centrifugal advance mechanism. A capacitor to reduce the arcing of the contact points was sometimes in the assembly and sometimes outside it.

Generally, the distributor was driven from a gear on the camshaft at half crankshaft speed. The ignition coil was normally mounted external to the distributor, but as close to it as possible to minimize the length of the high-voltage wire to the distributor cap.

The distributor built by Westinghouse Electric Co. in the early 1920s had the coil built into the distributor, but this was not common practice at the time. Ford continued to use magneto and vibrator coils throughout most of the 1920s, but changed to the breaker-type ignition system with the introduction of the Model A vehicle in 1928.

Batteries
During this period, the major development in batteries was the hard rubber battery case, which replaced one of wood. There were continuing improvements in performance as well. Delco Remy began the production of batteries in 1928. The period between 1912 to 1930 was a time of very rapid development of automotive technology. The fundamental concepts of the electrical system were well established during this time.

Hard-rubber-case battery.

1930 to 1960

Starters
Between 1930 and 1960, starter design remained relatively static. There were improvements in performance to match the changes in starting requirements as engine displacements changed. There were also improvements in durability, brush life, etc., but there were no major changes in concept.

Even the change made from 6 to 12 volts in the late 1950s had little impact on the starter. Because the starter could be greatly overpowered for its size due to its short operating time, when the change was made to 12 volts, it was overpowered even more.

Several issues were involved in the move to 12 volts. Engines were increasing in displacement, and V8s were becoming more common, requiring more cranking power. The first 12-volt systems were introduced by General Motors in 1955, and other manufacturers soon followed.

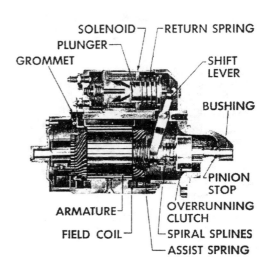

Solenoid-actuated starter motor.

The two methods of engagement—with a mechanical lever and the Bendix Drive—remained the most prominent. In 1956, Delco Remy introduced a solenoid-actuated electromechanical engagement mechanism with a totally enclosed shift lever—a concept which eventually came into widespread use.

By 1935 the number of suppliers had been reduced significantly. Autolite and Delco Remy supplied approximately 90% of all cars. Ford continued to supply its own requirements.

Generators

The designs for generators also stabilized during this period. They all were driven by a belt and had cutouts to prevent battery rundown. The third brush type of current limiting was common, and both current and voltage regulation were frequently used. Eventually a "three-unit regulator" was developed in which the cutout, current-sensing, and voltage-sensing relays were assembled into a single package. The temperature compensation of voltage for battery charging generally was implemented in these regulators by use of a bimetallic hinge for the regulator relay armature to change the

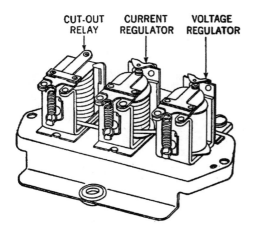

CUT-OUT RELAY   CURRENT REGULATOR   VOLTAGE REGULATOR

Typical vibrating-contact three-unit regulator.

magnetic force required to move the armature as the temperature changed. Most generators had a centrifugal cooling fan attached to the shaft behind the drive pulley. The fan drew air from the commutator end through the machine, and the air exited at the drive end.

In the late 1950s, when the change was made from 6 to 12 volts, the armature and field windings of the generator had to be changed to use twice the number of turns of wire and smaller wire size to match the 12-volt requirements. Beyond that, the basic design of the generator was not impacted significantly.

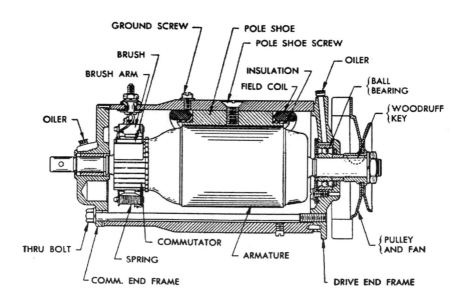

Typical dc generator.

Ignition

The Kettering ignition system had become the predominant system for practically all cars by this period. Magnetos had essentially disappeared. The distributor assembly did not change much during this time. It contained the breaker points, centrifugal and vacuum spark-advance mechanisms, the high-voltage distribution switch, and the capacitor.

Ford developed a unique distributor for use on its newly introduced 1932 V8 engines. It was attached to the front of the engine and was driven directly from the end of the camshaft. It had a high-voltage distribution cap on each side of the assembly, one for each bank of four cylinders. The ignition coil was integrally mounted to the top of the assembly. This distributor was unique to Ford.

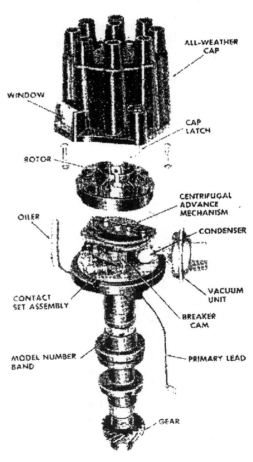

Reduction in maintenance was a continuing goal, and one method of extending the maintenance interval was to select material for the rubbing block on the breaker point arm that interfaced with the cam, such that its wear rate matched as closely as possible the erosion rate of the contact points. Thus the dwell adjustment would remain nearly constant as they wore together.

Radios in cars were first installed in the early 1920s. As their use increased, Radio

Delco Remy late-1950s distributor.

173

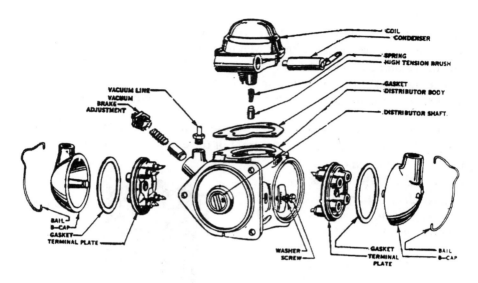

Ford V8 distributor.

Frequency Interference (RFI) became an increasing problem. Resistors were developed to be attached to the spark plugs, in series with the ignition wires, to suppress the noise. Eventually the conductor used in the ignition wire was designed to have the needed resistance, and separate resistors were no longer used. Also, resistors were built into the spark plug, which brought the RFI under acceptable control for that time.

The move to 12 volts benefited the ignition system because when cranking the engine with 6 volts, the voltage would drop well below 6 volts, significantly reducing the energy available to fire the spark plug. When cranking at 12 volts, the voltage would drop but adequate voltage remained to provide the desired spark energy. The move to 12 volts created a need for some modification of the ignition circuit. The 6-volt coil was retained, and a voltage-dropping resistor was used in series with the coil primary to drop the 12 volts to 6 volts under normal running conditions. However, when cranking, the resistor was shorted with a switch, either in the ignition switch or in the starter solenoid, so full battery voltage (8 to 11 volts) was applied to the coil. This provided considerably more spark energy during cranking than was available in the running mode with the 6-volt system. Thus, the move to 12 volts benefited both cranking operation and the available ignition energy.

Batteries

Basic battery construction did not change appreciably in this period. Hard rubber cases were used by all manufacturers. Battery repair became less fre-

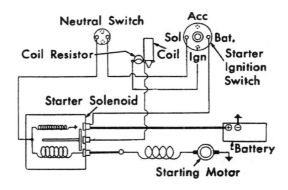

Typical ignition circuit showing the coil voltage-dropping resistor for the 12-volt system.

quent, and a significant replacement market developed. The move to 12 volts probably had its biggest impact on the battery. Batteries were now built with six cells instead of three. Plate area and number could be reduced because the current delivered was now half of what it had

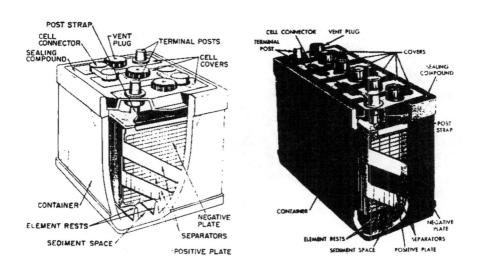

Comparison of 6- and 12-volt batteries.

previously been. The shape of the battery cases tended to become longer and narrower, reflecting the increased number of cells and the reduced plate area.

Overall, electrical system development during this time was focused on improvements to existing hardware designs. The change from 6 to 12 volts was the probably the most significant development for the period.

1960 to 1980

Starters

During this period, the use of electromechanical engagement of the starter pinion to the ring gear increased significantly. The solenoid-operated engagement mechanism was most common. Ford, however, developed a mechanism in which one of the pole shoes in the motor was hinged at one end, and the free end could move radially outward. The shoe was spring loaded in the outward position, and a linkage to the shift mechanism kept the pinion disengaged from the ring gear. When power was applied to the motor, the magnetic force, resulting from the current in the windings, pulled the pole shoe into its normal running position, engaging the pinion to the ring gear in the process.

During the 1970s, Chrysler began using gear reduction in its starters. This permitted the motor to develop its power at a higher shaft speed, which allowed it to be smaller and lighter. Chrysler had previously used a Delco

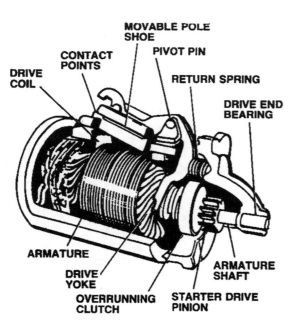

Ford starter with movable pole-shoe engagement.

Remy gear-reduction starter around 1930 but not in the intervening years. The geartrain in the starter gave a distinctive sound when cranking the engine, which distinguished Chrysler cars from others. Interestingly enough, the sound was sufficiently distinctive that sound-effects staff in the movies and on television used it when cars in their films were started,

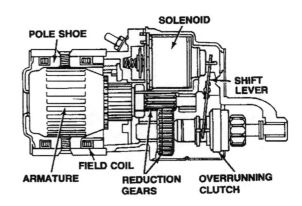

Chrysler gear reduction starter.

although cars other than Chryslers sometimes were being started. Since 1980, little has changed in starter technology with the exception of downsizing the motors as engines were downsized and friction was lowered.

Generators

During this period, generator developments produced a major change in vehicle-charging systems. The fundamental design of the electrical machine was totally changed, going from a common dc generator with a commutator to a Lundell type electromagnetic design. It used semiconductor diodes to rectify the ac to dc current. The diode bridge performed the same function as the commutator had in the dc machine. The shift to diode-rectified generators actually

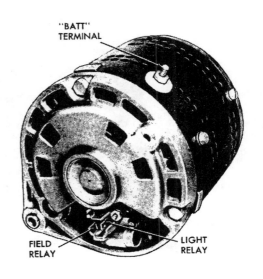

Delco Remy 1958 diode-rectified generator.

177

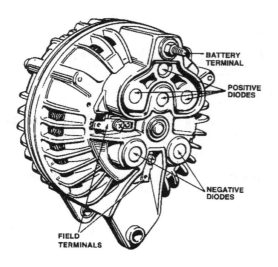

Chrysler diode-rectified generator.

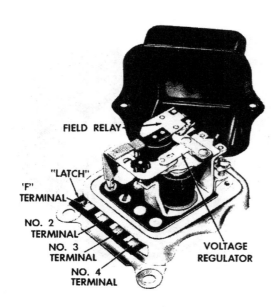

Typical two-unit regulator.

began in 1958 when Delco Remy introduced such a machine in applications with high electrical loads. Chrysler was the first to use them in high volume on passenger cars beginning in 1961, followed by GM in 1963 and Ford in 1965.

The driving force behind the change was the increasing need for electric power on cars. In 1960, power requirement was typically 500 watts. By the end of the 1980s, the power requirement on "up-option" cars was as high as 1500 watts, as more electrically powered devices were installed on vehicles.

The diode-rectifier bridge provided inherent protection against battery rundown because the diodes blocked any current that would flow from the battery through the generator; therefore, the cutout relay used with conventional dc generators was no longer needed. However, the field winding of the machine could run down the battery, and a field relay was substituted for the cutout relay. Regulation was still needed, and the same type of vibrating contact voltage regulation used with dc generators continued to be used initially. Instead of three-unit regulators, two-unit types were used. Current regulation was not needed because the generator's design was inherently current limited.

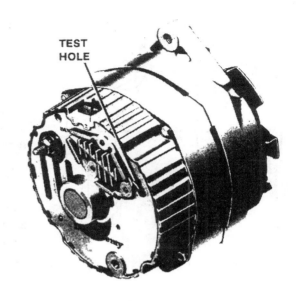

Delco Remy 10SI generator with integral regulator.

The first electronic voltage regulator was introduced with the high-power diode-rectified generator by Delco Remy in 1958. By the mid 1960s, electronic voltage regulators were in use on most cars. In 1968, Delco Remy introduced an optional generator with the electronic regulator built into the machine assembly. This has since become common practice.

The new generating system not only increased the power available on the vehicle but also significantly reduced maintenance. The brushes now had to carry field current of only 5 to 6 A, compared with brushes carrying full generator-output current in the dc generator. This greatly extended brush life. The electronic regulator eliminated the need for maintenance of the vibrating contacts in the electromagnetic regulators.

Ignition

Electronics found their way into many functions of the vehicle during this period, and ignition was no exception. Development began in the early 1960s to replace the contact points with a transistor switch. This required a transistor with high-voltage capability because transients could reach

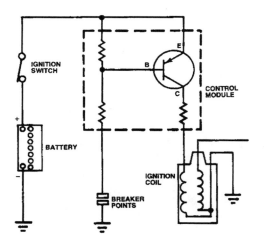

Contact-triggered electronic ignition circuit.

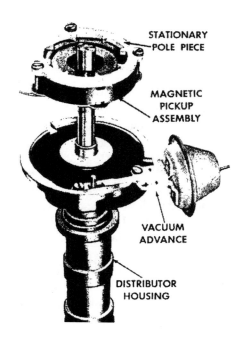

Delco Remy 1963 distributor with magnetic pickup.

300 volts in the ignition-coil primary winding. The early transistorized systems used a power transistor to switch the primary coil current, but retained the contact points as the triggering and synchronization device.

In 1963, the first magnetic pulse pickup was introduced by Delco Remy and was offered as an option on selected Pontiac engines that year. This magnetic pickup replaced the contacts as the triggering mechanism to operate the transistor switch. The use of electronic ignition increased through the 1960s and 1970s. By 1975, most vehicles produced in the United States used magnetic pickups.

In 1974, Delco Remy introduced its High Energy Ignition (HEI) system, with the coil in the distributor cap, on V8 engines. The HEI became the standard ignition system for GM cars. As a forerunner of the future, Delco Remy introduced in 1977 the first electronic ignition system in which spark timing was controlled with a digital microprocessor. It was used in the 1977 Oldsmobile Toronado.

The system replaced the functions performed by the centrifugal and vacuum advance mechanisms in earlier distributors.

In 1978, Ford introduced a microprocessor-based system to control spark timing and Exhaust Gas Recirculation (EGR), and thus began the proliferation of the use of microprocessors to control numerous functions of the car. Their increased use has been related to the need for more precise control of engine functions to meet tightening emissions and fuel economy requirements.

Batteries

In the mid 1960s, Johnson Controls introduced a battery case made of polypropylene. The conventional hard rubber was replaced by this new plastic case material, which made the case lighter and less susceptible to cracking. The material has become an industry standard.

Battery designs, except for the new case material, had not changed significantly until 1969, when Delco Remy introduced the maintenance-free battery. It required no addition of water to the electrolyte during its lifetime. The vent caps were eliminated, with venting achieved by use of a small port at the top of the case.

Delco Remy maintenance-free battery.

Several developments allowed the battery to be maintenance free. Among them were the use of calcium to replace the antimony in the lead alloy grid, the development of "microporous" polyethylene material for the plate separators, and the increased volume of the electrolyte reservoir. The calcium reduced the consumption of water in the electrochemical process of charging and discharging. The use of the porous plastic separator reduced the tendency for internal shorting with lead-calcium grids, and the increased volume of electrolyte was large enough that no water had to be added during the life of the battery. The developments of this period were to shape the battery business for many years to come.

1980 to the Present

Starters

Starter technology did not change much from the previous period. Gear-reduction motors came into more widespread use, and in 1983 Delco Remy introduced a starter with permanent magnets to replace the field coils, in a gear reduction configuration. It was used on some light-duty trucks but in limited volume.

Generators

Generators during this period did not experience major developments. There was a continuing effort to reduce size and weight, improve life, and reduce noise. As engine developments resulted in lower noise levels, pressure increased on noise reduction for the other components in the engine compartment as well, including the generator.

Ignition

The major tightening of emissions and fuel economy standards in 1981 brought significant changes in ignition and fuel control systems. They can no longer be viewed as separate entities because the microprocessor controller, which manages the control of engine parameters (ECM), combines the two functions interactively. The ECM controls exhaust gas recirculation, idle speed, and the air injection system as well. Instead of ignition or fuel control, the discussion must now focus on engine management systems.

To perform its control functions, the ECM had to have a great deal of information about engine operating parameters. A large number of transducers were needed to convert engine physical parameters into electrical signals suitable for use by the ECM. Among the sensors developed for the new systems were crankshaft and camshaft speed and angular position sensors, manifold absolute pressure and other pressure sensors, throttle position sensors, and temperature sensors. An oxygen sensor in the exhaust stream also was needed to determine if the air/fuel ratio of the engine was near stoichiometry.

Fuel is now delivered to the engine through electromagnetic fuel injectors, providing more precise fuel metering than was possible with carburetors. The distributor has been reduced to a high-voltage switch to provide energy to the spark plugs in the proper firing order.

In recent years, the distributor on many engines has been eliminated entirely in favor of distributorless systems which use multiple coils. In these designs, each of the two ends of the coil secondary winding are connected to spark plugs. The two cylinders, whose spark plugs are attached to one coil, are selected so that when one is on its compression stroke ready to be fired, the other is on its exhaust stroke. Its spark plug fires at the same time as the one on the compression stroke, but there is nothing to burn in that cylinder and its spark has no effect.

The precision with which the engine operation can be controlled by these systems has been significantly improved, but more precise control will be needed to achieve further reductions in emissions and fuel consumption. The complexity of these engine management systems is such that space does not permit a description of their operation; however, a great deal of literature is currently available covering the subject.

The microprocessor controller enabled improved control of engine systems, and the technology has now been applied to other functions of the car. Anti-lock brakes, functions within the body and chassis, and entertainment systems now have microprocessors involved in their operation.

It is apparent that the automobile electrical system has come a long way in its development in the hundred years since the first sparks were created to ignite the air/fuel mixture in the cylinders of the early engines. There is little reason to think that it will not continue to change in the future.

What About the Future?

Power requirements on cars continue to increase as more electrically operated accessories are contemplated. These include items such as electrically heated catalysts, electric power steering, electric air conditioning compressors, more sophisticated suspension systems, and other loads that will be added to the vehicle. As the requirements for further reduction in emissions and fuel consumption are contemplated, higher efficiencies will be required throughout the electrical system. Also, it is apparent from recent developments that the electrical system will likely become a participant in the propulsion of the vehicle.

At recent auto shows, several systems have appeared in which high-powered electrical systems have been integrated into the propulsion and energy recovery systems of the vehicles. Instead of electrical systems of 1 or 2 kW, these systems will have the capability of developing 20, 30, or 50 kW. The electric vehicle, of course, uses electric power for all of its operation and propulsion, as well as heating, cooling, and accessory functions.

The role of electricity on the car will continue to expand in the future. The technology is available today to meet these needs, but the challenge will be to provide the customer with a cost-effective system that will meet the needs of future vehicles.

Gasoline Specifications, Regulations, and Properties

Lewis M. Gibbs
Chevron Research and Technology Company

Gasoline was used as the fuel for the first four-stroke (Otto) cycle spark-ignition engine around 1884, and at that time it was considered an undesirable by-product of kerosene manufacturing. As the demand for gasoline increased from 1900 to 1920, it ceased to be a by-product and the more volatile fractions of kerosene were diverted to gasoline. Thermal cracking in 1913 and other refining processes in subsequent years were introduced to convert a larger fraction of petroleum to gasoline.

Federal Specifications

The first gasoline specification that appeared in the literature was a federal specification issued in 1919. The only properties it specified were maximum limits on the distillation initial boiling point; the 20% evaporated, 50% evaporated, and 90% evaporated points; and end point. The limits of

Table 12-1
Gasoline Specifications

Property	1919 Federal Limits	1929 Federal Limits	1937 ASTM D 439-37T Note 1	1970 ASTM D 439-70	1988 ASTM D 4814-88
Distillation, °C (°F)					
Initial BP	60 (140) Max.	-	-	-	-
10% Evaporated	-	50-80 (122-176)	-	-	-
Class W	-	-	60 (140) Max.	-	-
Class F	-	-	65 (149)	-	-
Class S	-	-	70 (158) Max.	-	-
Class A	-	-	-	70 (158) Max.	70 (158) Max.
Class B	-	-	-	65 (149) Max.	65 (149) Max.
Class C	-	-	-	60 (140) Max.	60 (140) Max.
Class D	-	-	-	55 (131) Max.	55 (131) Max.
Class E	-	-	-	50 (122) Max.	50 (122) Max.
20% Evaporated	105 (221) Max.	-	-	-	-
50% Evaporated	140 (284) Max.	140 (284) Max.	140 (284) Max.		
Class A	-	-	-	77-121 (170-250)	77-121 (170-250)
Class B	-	-	-	77-118 (170-245)	77-118 (170-245)
Class C	-	-	-	77-116 (170-240)	77-116 (170-240)
Class D	-	-	-	77-113 (170-235)	77-113 (170-235)
Class E	-	-	-	77-110 (170-230)	77-110 (170-230)
90% Evaporated	187 (369) Max.	187 (369) Max.	200 (392) Max.		
Class A	-	-	-	190 (374) Max.	190 (374) Max.
Class B	-	-	-	190 (374) Max.	190 (374) Max.
Class C	-	-	-	185 (365) Max.	185 (365) Max.
Class D	-	-	-	185 (365) Max.	185 (365) Max.
Class E	-	-	-	185 (365) Max.	185 (365) Max.
End Point	225 (437) Max.	225 (437) Max.	-	225 (437) Max.	225 (437) Max.
Vapor Pressure, kPa (psi)					
Class W	-	-	93 (13.5)	-	-
Class F	-	-	79 (11.5)	-	-
Class S	-	-	65.5 (9.5)	-	-
Class A	-	-	-	62 (9.0)	62 (9.0)
Class B	-	-	-	69 (10.0)	69 (10.0)
Class C	-	-	-	79 (11.5)	79 (11.5)
Class D	-	-	-	93 (13.5)	93 (13.5)
Class E	-	-	-	103 (15.0)	103 (15.0)
Gum, mg/100 mL	-	-	7 Max.	5 Max.	5 Max.
Corrosion	-	-	Passes	1A	1A
Motor Octane Number	-	-	67R/75P Min.	83R/88P Min.	
Research Octane Number	-	-	-	90R/96P Min.	-
Temp. for V/L=20, °C (°F)					
Class A	-	-	-	60 (140) Min.	60 (140) Min.
Class B	-	-	-	56 (133) Min.	56 (133) Min.
Class C	-	-	-	51 (124) Min.	51 (124) Min.
Class D	-	-	-	47 (116) Min.	47 (116) Min.
Class E	-	-	-	41 (105) Min.	41 (105) Min.
Oxidation Stability, minute	-	-	-	-	240 Min.
Sulfur Content, mass %	-	-	-	-	0.15 Pb/0.10UL Max.
Lead Content, g/L (g/gal)					
Leaded	-	-	-	-	1.1 (4.2) Max.
Unleaded	-	-	-	-	0.013 (0.05) Max.
Water Tolerance	-	-	-	-	Area Table

Note 1: Type A shown, Type B 50% evaporated maximum 125°C (257°F) and 90% evaporated maximum 180°C (356°F)

the 1919 federal specification are presented in Table 12-1. In 1929, the federal specification dropped the limits for initial boiling point and the 20% evaporated point in favor of a range for the 10% evaporated point.

ASTM Specifications

The first consensus gasoline standard issued by the American Society for Testing and Materials (ASTM) appeared in a tentative specification in 1937 as ASTM D 439-37T. A tentative specification was not a full consensus standard and had a life of only three years. However, it could have been extended when the intent was to make it a standard specification. Tentative specifications are no longer part of ASTM. The limits shown in Table 12-1 are for Type A gasoline. A more volatile grade, Type B, with different 50% evaporated and 90% evaporated points, is covered by Note 1 of Table 12-1. There also is a relatively nonvolatile Type C gasoline with the same 50% evaporated and 90% evaporated limits as Type A, but with a fixed 10% evaporated maximum of 75°C (167°F). Within Types A and B, there were three volatility classes—a winter class W, a transition class F, and a summer class S. Only the 10% evaporated point and vapor pressure maximum limits changed with the seasons. This was the first approach to climatized gasoline. In addition to volatility limits, there were maximum gum content, corrosion, and minimum Motor octane number limits (for regular- and premium-grade gasolines). Research octane number was not specified because it did not become an ASTM tentative test method until 1947. D 439-37T evolved with time and became a standard specification in 1970 (ASTM D 439-70) as shown in Table 12-1. By this time, there were five volatility classes (A–E) for which the maximum 10% evaporated point, range in 50% evaporated point, maximum 90% evaporated point, maximum end point, maximum vapor pressure, and minimum temperature for a vapor-liquid ratio of 20 were changed with the seasons. The volatility class is specified by state or portion of state for each month of the year. Minimum Research and Motor octane numbers were specified for regular- and premium-grade gasolines but were negotiable, especially for high-altitude areas. The gum content limit had been reduced, and a specific corrosion limit was required.

Over the years, this specification evolved with the addition of maximum sulfur content limits (1971, 1973, and 1977), maximum lead content limits for both leaded and unleaded gasoline (1971 and 1975), minimum oxidation stability (1979), replacement of octane limits with an Antiknock Index applications table (1971), and Antiknock Index adjustments for various altitudes. It was replaced in 1988 by D 4814 Standard Specification for Automotive Spark-Ignition Engine Fuel. This new specification covered gasoline-oxygenated blends and gasoline, and its requirements are summarized in Table 12-1. The latest version available is ASTM D 4814-95c and now provides an alphanumeric system for specifying volatility requirements. The alpha part specifies vapor pressure and distillation limits; the numeric part sets minimum temperatures for a vapor-liquid ratio of 20. The current version complies with federal Phase II conventional gasoline vapor pressure regulations, and modifications are underway to comply with the federal reformulated gasoline (RFG) simple model vapor pressure rules.

Air Pollution Regulations

The first regulation designed to help reduce air pollution was issued in 1959 by the Los Angeles County Board of Supervisors. It set a maximum bromine number limit of 30 to control the amount of olefins in Los Angeles County gasoline as the first step to limit the formation of eye irritants, ozone, and aerosols. In 1971, this regulation was taken over by the California Air Resources Board (CARB) and applied to South Coast Air Basin gasoline. In that same year, CARB implemented the "nine-pound rule," which limited vapor pressure in California to 62.0 kPa maximum for the summer months (the time period depended on the air basin) as a first step to reduce evaporative hydrocarbons.

The first federal requirement related to air pollution became effective in July 1974 when one grade of unleaded gasoline having a minimum Research octane number of 91 was required in nearly all service stations. The fuel also could contain no more than 0.013 g/L lead and no more than 0.0013 g/L phosphorus. The U.S. Environmental Protection Agency (EPA) wanted to ensure a supply of the proper fuel for 1975 model catalyst-equipped vehicles.

Table 12-2
Summary of Gasoline Related Air Pollution Regulations

Year	Agency	Regulation	Purpose
1959	CA	Bromine Number–30 Max. for Southern CA	To limit formation of eye irritants, ozone, and aerosols
1971	CA	Vapor Pressure–9.0 psi Max. summer months	To reduce evaporative hydrocarbon emissions and ozone
1974	US	Unleaded Gasoline Required in Service Stations	To assure proper fuel for exhaust catalyst-equipped vehicles
1976	CA	Sulfur Limit–500 ppm Max.	Reduce sulfur dioxide and sulfur trioxide (sulfate emissions)
1977	CA	Lead Phasedown	To protect public health
1977	US	Manganese Banned Until Waiver Obtained	To prevent increase in hydrocarbon emissions
1977	CA	Manganese Banned	To prevent increase in hydrocarbon emissions
1978	CA	Sulfur Limit–400 ppm Max.	Reduce sulfur dioxide and sulfur trioxide (sulfate emissions)
1980	US	Lead Phasedown	To protect public health
1980	CA	Sulfur Limit–300 ppm Max.	Reduce sulfur dioxide and sulfur trioxide (sulfate emissions)
1981	US	Substantially Similar Rule	To control additive and oxygenate use
1989	US	Vapor Pressure Phase I 10.5/9.5/9.0 psi Max. Summertime	To reduce evaporative hydrocarbon emissions and ozone
1992	US	Vapor Pressure Phase II 9.0/7.8 psi Max. Summertime	To reduce evaporative hydrocarbon emissions and ozone
1992	US	Oxygen Content–2.7 wt % Min. Wintertime administered by states	To reduce wintertime carbon monoxide emissions in carbon monoxide nonattainment areas
1992	CA	Vapor Pressure Phase 1 7.8 psi Max. Summertime	To reduce evaporative hydrocarbon emissions and ozone
1992	CA	Deposit Control Additive Requirement	To minimize exhaust emissions caused by carburetor, injector, and intake valve deposits
1992	CA	Lead Banned	To protect public health
1992	CA	Oxygen Content–1.8-2.2 wt % Wintertime	To reduce carbon monoxide wintertime emissions without increasing oxides of nitrogen emissions
1994	CA	Required All Gasoline to be Unleaded	To protect public health and catalysts
1995	US	Deposit Control Additive Requirement	To minimize exhaust emissions caused by carburetor, injector, and intake valve deposits
1995	US	Reformulated Gasoline Simple Model	To reduce ozone in specified and opt in ozone nonattainment areas
		Benzene Limit–1.3 vol % Max. per gallon cap	To reduce toxics
		Oxygen Content–1.5 Wt % Min. per gallon cap	Required by CAAA 1990
		Vapor Pressure–7.4/8.3 psi Max. per gallon cap	To reduce evaporative hydrocarbon emissions and ozone
		No Heavy Metals	To protect public health
		Indirect Aromatics Max. of ~27 vol %	To reduce toxics
		Sulfur, olefins, and 90% evaporated point <1990 average levels	To prevent increased emissions caused by changes in other fuel properties
1996	US	Lead Banned for Highway Fuel	To protect public health
1996	CA	California Phase 2 Reformulated Gasoline	To achieve maximum cost-effective reductions in criteria and toxic pollutants
		Vapor Pressure–7.00 psi Max.	To reduce evaporative hydrocarbon emissions and ozone
		Sulfur Limit–80 ppm Max. per gallon cap	Reduce sulfur dioxide and sulfur trioxide (sulfate emissions) and minimize temporary deactivation of exhaust catalysts thereby reducing hydrocarbon, carbon monoxide, and oxides of nitrogen emissions
		Benzene Limit–1.2 vol % Max. per gallon cap	To reduce toxics
		Aromatics Limit–30 vol % Max. per gallon cap	To reduce toxics and hydrocarbon emissions
		Olefins Limit–10.0 vol % Max. per gallon cap	To reduce oxides of nitrogen exhaust emissions and ozone formation from evaporative emissions
		90% Evaporated Point–330°F Max. per gallon cap	To reduce hydrocarbon exhaust emissions
		50% Evaporated Point–220°F Max. per gallon cap	To reduce hydrocarbon and carbon monoxide emissions
		Oxygen Content–0-2.7 Wt % Summertime	To reduce carbon monoxide and hydrocarbon emissions without increasing oxides of nitrogen emissions

Table 12-2 summarizes the California and federal limits and the purpose of the various regulations, beginning with the first regulation in 1959 to the 1996 California Phase 2 reformulated gasoline (RFG) requirements. The regulations control the amount of sulfur in gasoline (California—1976, 1978, 1980, and 1996), phase down the lead content of leaded gasoline (California—1977; federal—1980), and finally eliminate leaded gasoline (California—1992; federal highway—1996). They also reduce vapor pressure several times (California—1971, 1992, and 1996; federal—1989, 1992, and 1995) and require the use of oxygenates (California—1992 and 1996; federal—1992 and 1995). Regulations require the use of deposit control additives to prevent port fuel injection and intake valve deposits (California—1992; federal—1995), limit benzene content (California—1996; federal—1995), control hydrocarbon composition (California—1996), and limit the maximum 50% and 90% evaporated points (California—1996). The federal RFG simple model requirements for 1995, which apply only to specified and opt-in ozone nonattainment areas in the United States, indirectly limit aromatics to control toxics and set a cap on sulfur and olefins contents and the 90% evaporated point based on 1990 refinery averages.

The federal RFG regulations require that the reformulated fuel be used in specified severe ozone nonattainment areas. Other ozone nonattainment areas may join the program as opt-in areas, and many have. Now some opt-in areas have decided to opt out of the RFG program. At this time, no procedure exists for opting out; however, one has been proposed by EPA, and enforcement is on hold for those areas that have requested to opt out.

Impact of Specifications and Regulations
Vapor Pressure
Although distillation properties were the first properties controlled by specifications, vapor pressure specifications and vapor pressure reduction regulations have had a greater influence on gasoline properties. Figure 12-1 shows how the U.S. national average summer and winter vapor pressures for both regular- and premium-grade gasoline have changed from 1955 (before any air-pollution-related regulations came into effect) through 1995. For 1995, data are shown for both conventional gasoline and RFG. Indicated on Figure 12-1 are the times that ASTM D 439 (1970) was

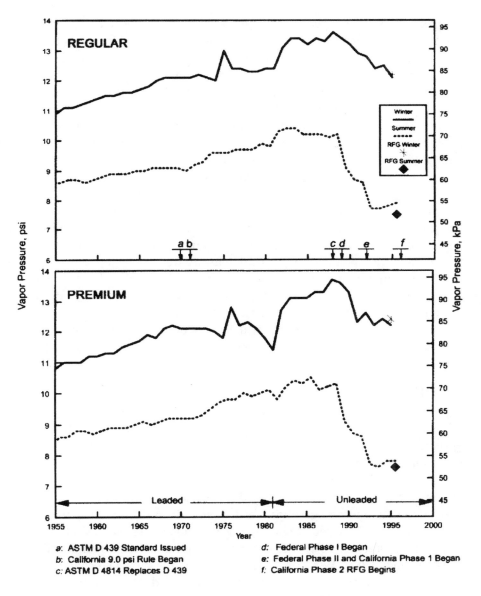

a: ASTM D 439 Standard Issued
b: California 9.0 psi Rule Began
c: ASTM D 4814 Replaces D 439

d: Federal Phase I Began
e: Federal Phase II and California Phase 1 Began
f: California Phase 2 RFG Begins

Fig. 12-1. U.S. national average vapor pressure trends.

adopted as a standard specification and when it was replaced by D 4814 (1988), when the various California maximum vapor pressure limits became effective (1971, 1992, and 1996), and when the federal Phase I (1989) and Phase II (1992) vapor pressure regulations began. Until 1981, the data are shown for leaded gasoline. After 1981, unleaded gasoline data are shown because it was now the predominant fuel. For both summer and winter gasolines, an upward trend in vapor pressure continued until 1989, at which time there was a sudden drop for summer gasolines in response to federal Phase I vapor pressure regulations. The reason for the upward trend was the increasing amount of butane and other light hydrocarbons from refinery processing that were striving to meet octane and volume demands as unleaded gasoline sales increased and lead content was being phased down in leaded gasoline. Surprisingly, the winter gasoline also showed a downward trend after 1989, although there are no federal winter vapor pressure controls. The downward trend in vapor pressure continued with the implementation of the federal Phase II and California Phase 1 regulations until 1992, when it appeared to level out. Also shown in Figure 12-1 are the average data for RFG. The winter gasoline vapor pressure does not differ from that of the conventional gasoline, but the summer gasoline is slightly lower for both regular- and premium-grade gasolines.

Distillation

Figure 12-2 is a histogram showing the trend with time for the U.S. national average 10% evaporated point, 50% evaporated point, and 90% evaporated point for both regular- and premium-grade gasolines. The trend has been downward from 1955 to 1981 for leaded gasoline for the 10% and 50% evaporated points. An increase is shown for the 50% evaporated point when the data base was changed from leaded to unleaded gasoline, indicating that unleaded gasoline was less volatile in the midrange than leaded gasoline. Unleaded regular-grade gasoline has shown a slight increase in the summer 90% evaporated point with time but has remained constant since 1988. Winter unleaded regular-grade shows no change. There has been no significant change in the unleaded premium-grade 90% evaporated point. For both grades, there has been a rise in the summer 10% evaporated point, which corresponds to the implementation in 1989 of the federal Phase I vapor pressure regulation reductions, and no significant increase in the 50% evaporated point. For winter, the 10% evaporated

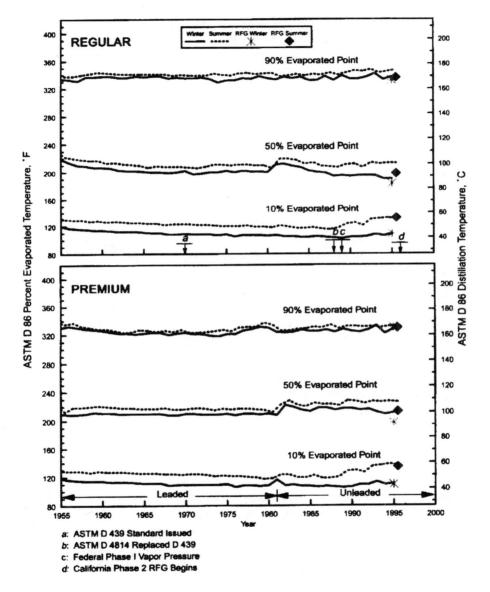

Fig. 12-2. U.S. national average distillation trends.

a: ASTM D 439 Standard Issued
b: ASTM D 4814 Replaced D 439
c: Federal Phase I Vapor Pressure
d: California Phase 2 RFG Begins

point increased in response to the observed decrease in vapor pressure shown in Figure 12-1, while the 50% evaporated point continued its decline. The increase in the 10% evaporated point was caused by the removal of butane to lower vapor pressure. There was no difference in the 10% evaporated point between RFG and conventional gasoline for both grades. The 50% evaporated point was lower for RFG for both winter and summer gasolines for both grades. The only difference in the 90% evaporated point occurred for regular-grade gasoline.

Driveability Index

One calculated property, Driveability Index (DI = 1.5 x 10% evaporated temperature + 3.0 x 50% evaporated temperature + 1.0 x 90% evaporated temperature), is not a specification limit in ASTM D 4814, but its adoption has been seriously considered several times. Driveability is an assessment of a vehicle's response to the driver's movement of the accelerator pedal. Low values of DI are considered desirable, but less satisfactory performance for some cars occurs only when some threshold level is exceeded. Figure 12-3 shows that U.S. national average DI became lower with time for both leaded regular- and premium-grade gasolines. Unleaded gasoline grades had higher DIs than the leaded gasolines they replaced because heavier aromatic hydrocarbons were used to provide antiknock performance in place of lead antiknocks. Winter unleaded regular-grade DI has improved with time; the premium-grade remained unchanged. Summer unleaded gasoline DIs have increased as a result of the regulations requiring the reduction of vapor pressure. Inspection of Figure 12-2 to determine what distillation factor had the largest effect on the increase shows that it was the 10% evaporated point increase. The data show that refiners did not add heavier components to make up for the volume that was lost because of butane removal to meet the reduced vapor pressure limits. If this had occurred, the 90% evaporated point would have increased. It would have been difficult for refiners to compensate for the increase in DI caused by the butane removal. This is because more high-boiling components would have to be removed than butane because the DI equation 10% evaporated parameter is more heavily weighted in the equation than the 90% evaporated parameter. This would cause an even larger loss of product volume and a serious disposal problem. The high-boiling fractions are high in aromatics content and low in flash point. Only limited amounts of this material could be used in aviation turbine and diesel fuels.

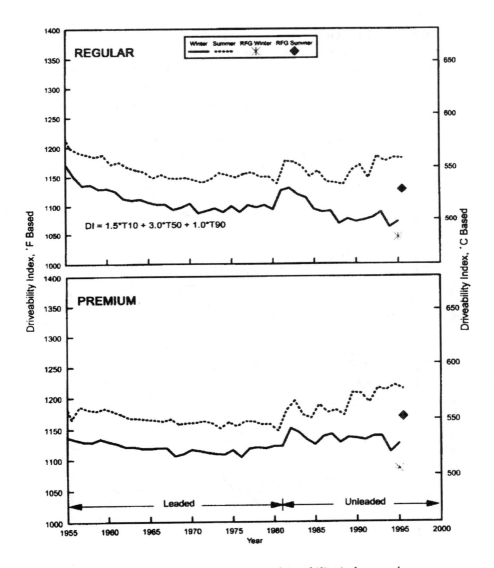

Fig. 12-3. U.S. national average driveability index trends.

New processing would be required by most refiners to dispose of the high-boiling gasoline fractions. For both winter and summer, the DIs are lower for RFG than for conventional fuels. This is primarily a result of the lower 50% evaporated points for RFG.

Vapor-Liquid Ratio

The minimum temperature for a vapor-liquid ratio of 20 is specified to protect against vapor lock. This is a property that has not been measured and reported in gasoline surveys. There are equations for calculating this property from vapor pressure and distillation properties; however, the

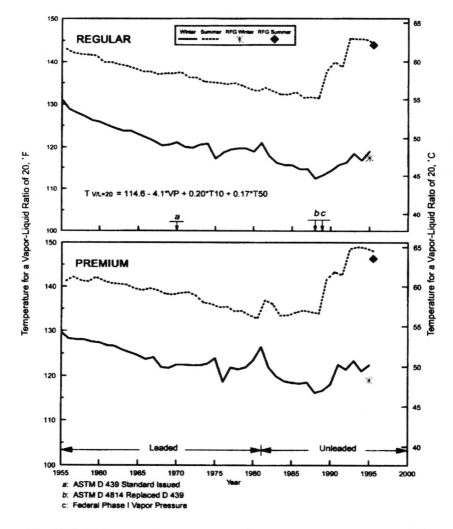

$$T_{V/L=20} = 114.6 - 4.1*VP + 0.20*T10 + 0.17*T50$$

a: ASTM D 439 Standard Issued
b: ASTM D 4814 Replaced D 439
c: Federal Phase I Vapor Pressure

Fig. 12-4. U.S. national average temperature for a vapor-liquid ratio of 20.

equations do not work for fuels containing ethanol, and there is a small bias for ether-containing fuels. Even so, an estimate of the trend can be shown by using the linear equation in ASTM D 4814. Figure 12-4 shows that the temperature for a vapor-liquid ratio of 20 was decreasing until the federal Phase I vapor pressure regulation took effect. Since then, the vapor-liquid ratio temperature for both summer and winter gasolines has increased rapidly. RFG has approximately the same values as conventional gasoline for regular-grade and slightly lower for premium-grade gasoline.

Lead Content

Another property that has been regulated is the lead content of leaded gasoline. As shown in Table 12-2, California began phasing down the lead content in 1977, and the federal phase-down program began in 1980. The purpose of these programs was to protect public health, particularly for children. Figure 12-5 shows the dramatic impact of the lead phase-down regulations on the lead content of leaded gasoline. The usage of lead

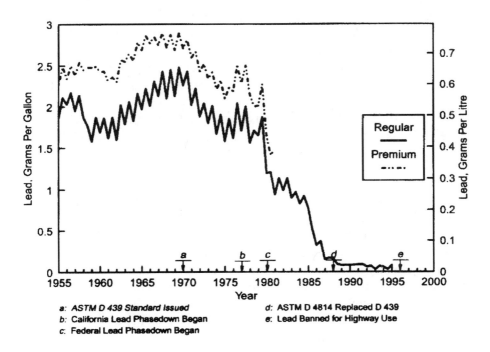

a: ASTM D 439 Standard Issued
b: California Lead Phasedown Began
c: Federal Lead Phasedown Began

d: ASTM D 4814 Replaced D 439
e: Lead Banned for Highway Use

Fig. 12-5. U.S. national average lead usage trends.

peaked around 1969, and leaded premium-grade gasoline essentially disappeared in 1981. Although the lead content limit was reduced to an average 0.026 g/L in 1986, the actual level was above the limit until 1988 because the regulations allowed for the banking of lead credits. Leaded gasoline was eliminated from California in 1992, and in 1996 lead was eliminated from highway gasoline by federal regulation.

Although the maximum federal lead content limit for unleaded gasoline is 0.013 g/L lead, the actual level averages below 0.0003 g/L. With the elimination of leaded gasoline for highway use in the United States in 1996, the lead level in unleaded gasoline will be below the level detectable by current standard gasoline test methods.

Antiknock Index

Octane quality of U.S. gasoline is represented by the average of the Research and Motor octane numbers, known as the Antiknock Index. While minimum octane limits have been part of some ASTM specifications, they were eliminated before the Federal Trade Commission required the posting of the Antiknock Index on gasoline dispensing pumps in 1979. Figure 12-6 presents the U.S. national trend in Antiknock Index for various leaded and unleaded gasoline grades. Antiknock Index peaked in the early 1970s and dropped during the 1973 oil embargo and the 1979 energy crisis. Unleaded premium-grade has shown a 1.2 number increase in Antiknock Index from 1981 to 1995. Unleaded regular-grade overall has not changed since 1981, as shown in Figure 12-6. The Antiknock Index of RFG does not differ from the corresponding conventional gasoline grades.

Sulfur Content

Only California has required a reduction in gasoline sulfur content. One sulfur reduction program began in 1976, and another became effective in 1996. Under the federal RFG simple model program, a sulfur cap based on 1990 refinery averages has been established for both RFG and conventional gasoline. Figure 12-7 shows that sulfur content has decreased over time, primarily as a result of the increased use of catalytic reformers and the hydrotreating of the feed to catalytic crackers. The unleaded gasoline grades initially had lower sulfur contents than the leaded grades. However, as demand increased for unleaded regular-grade and decreased

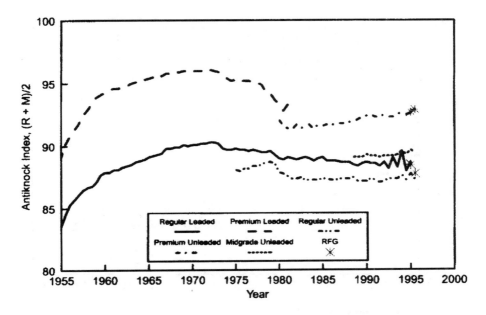

Fig. 12-6. U.S. national average antiknock trends.

for leaded regular-grade, the sulfur levels of the two products became comparable. The sulfur level of RFG is approximately the same for regular-grade and somewhat higher for premium-grade as compared to the conventional gasolines.

Aromatics Content

There are no limits on the amount of aromatics in ASTM D 4814, and the federal RFG simple model indirectly limits aromatics by its toxics reduction requirement. California Phase 2 RFG has set a maximum limit on aromatics content for 1996 to reduce toxics and hydrocarbon emissions. Figure 12-8 shows how the U.S. national average aromatics content has changed with time. Unleaded regular- and premium-grade gasolines have higher aromatics contents than the leaded grades they replaced. As lead was phased down, the aromatics content of leaded gasoline increased. The current trend shows a lowering of the aromatics content. Part of this reduction can be attributed to dilution caused by the increased use of

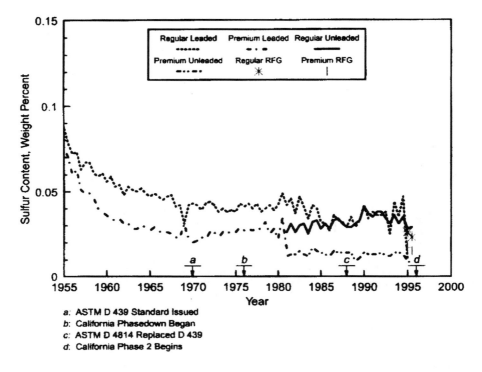

Fig. 12-7. U.S. national average sulfur content trends.

oxygenates for octane improvement and to meet federal oxygenated gasoline requirements. The RFG samples have lower aromatics contents than conventional gasoline, as shown in Figure 12-8.

Olefins Content

Los Angeles County first used a maximum bromine number limit to control olefins in gasoline to limit the formation of eye irritants, ozone, and aerosols. Now California Phase 2 RFG requirements have set a limit on olefins content to reduce oxides of nitrogen emissions and ozone formation from evaporative emissions. Under the federal RFG simple model program, an olefin cap based on 1990 refinery averages has been established for both RFG and conventional gasoline. Figure 12-9 shows the changes in U.S. national average olefins content with time. Since 1975, the olefins content from unleaded regular-grade has been increasing, while the level in

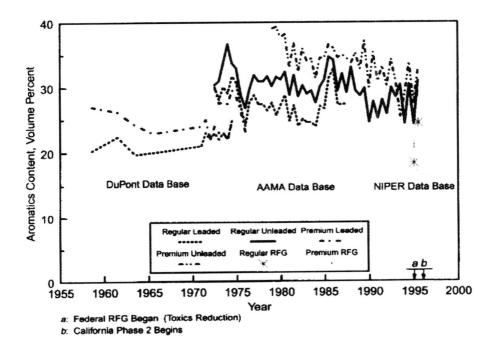

a: Federal RFG Began (Toxics Reduction)
b: California Phase 2 Begins

Fig. 12-8. U.S. national average aromatics content trends.

unleaded premium-grade has shown a slight downward trend. The olefins content of RFG is approximately the same as for conventional gasoline, although a reduction is expected when 1996 California data are available.

The Future

Both the federal and California RFG programs have a complex model or predictive model that contains formulas that allow trading the lowering of the level of one property for the increasing of the level of another property. Further, by the year 2000, the federal Phase II RFG requirements take effect and require additional emissions reductions. Concern has been expressed by automobile manufacturers that the performance of onboard diagnostic (OBD) II systems may be sensitive to sulfur content. As part of their federal State Implementation Plan (SIP) to meet ozone attainment, some states have set summer maximum vapor pressure limits below those required by the federal Phase II vapor pressure regulations. Some areas

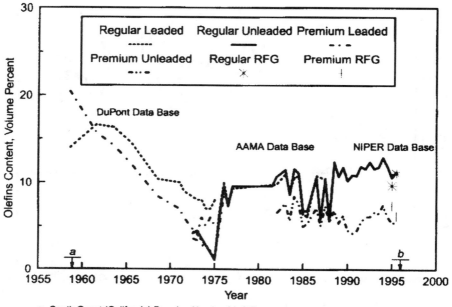

Fig. 12-9. U.S. national average olefins content trends.

even have set maximum vapor pressure limits for winter. Thus, gasoline properties in the future may approach those currently specified by the California Phase 2 RFG regulations.

For More Information

For gasoline property trends prior to 1955, please refer to SAE Paper 932828, "How Gasoline Has Changed," and to SAE Paper 902104, "Gasoline Additives—When and Why." For copies of the latest version of the D 4814 Standard Specification for Automotive Spark-Ignition Engine Fuel and for information on RFG in the ASTM RR: D02-1347 Research Report on Reformulated Spark-Ignition Engine Fuel, contact ASTM in West Conshohocken, Pennsylvania.

Chapter 13

A History of Lubrication

Syed Q.A. Rizvi
The Lubrizol Corporation

All mechanical equipment must be lubricated. The purpose is to reduce friction and wear. If not controlled, these can lead to inefficiencies, damage, and ultimately equipment seizure.

Pictorial records depicting the use of lubricants date as far back as 1650 B.C. Analysis of the residue from chariot axle hubs suggests the use of animal fat as a lubricant as early as 1400 B.C. This practice continued until 1859, when petroleum-based lubricants became available. Most modern lubricants are petroleum-based, although vegetable oils and animal fats are also used. All lubricants contain chemicals, called additives. Without their presence, effective lubrication of modern machinery is not possible. Common types of automotive lubricants include engine oils, transmission fluids, gear oils, and greases. Lubricant performance is generally defined by viscosity, service classifications, and/or specifications, and evolves with changes in hardware, design, and operating conditions.

Engine Oils

Engine oils are designed to reduce friction, minimize deposit formation, and prevent corrosion and wear. The quality of these oils is defined by the Society of Automotive Engineers (SAE) viscosity classifications, the

American Petroleum Institute (API) service classification system, and the specifications established by U.S. military and original equipment manufacturers (OEMs).

Viscosity Classifications

Viscosity is one of the most important properties of a lubricant. It can be defined as a lubricant's resistance to flow. Proper viscosity in a lubricant minimizes surface-to-surface contact, hence friction and wear. The importance of viscosity was recognized in the early part of this century when, in 1911, SAE established the first engine oil classification system based on viscosity. The 1923 revision included specifications for 10 oils that were classified according to their viscosity ranges. The specification numbers ranged from 20 to 115 and were based on the first two digits of the average Saybolt viscosity in seconds (SUS). The numbers between 20 and 50 were based on viscosity at 37°C (100°F), and the numbers between 60 and 115 were based on viscosity at 98.8°C (210°F). Two special specifications, 020 and 030, were assigned to oils with pour points $\leq$-12.2°C ($\leq$10°F). This classification was of limited use because marketing and owner's manual designations of "light," "medium," etc. did not properly fit its viscosity grades. Hence, in 1926, a classification system based on viscosities at 54.4°C (130°F) and 98.8°C (210°F) was established in which the classification numbers had no relationship to actual viscosities.

In 1933, the Lubricants Division tentatively introduced two W grades for which viscosity at -17.7°C (0°F) was determined by extrapolation from higher temperatures. In 1950, SAE grades 10, 60, 70, and the 54.4°C (130°F) requirements were dropped. A new grade 5W was added, and 10W and 20W were officially included in the classification. Multiviscosity grades became part of the classification in 1955, and the classification became official in 1962 as SAE J300. In 1967, the cold cranking method to determine -17.7°C (0°F) viscosity of the W grades replaced the extrapolation method, and kinematic viscosity to measure 98.8°C (210°F) viscosity became a primary method. In 1967 and 1974, additions were made to the classification without changing the number of viscosity grades. However, in 1975, a new grade 15W was added. There have been several revisions since then. The March 1993 revision, along with the most recent military requirements, is given in Tables 13-1a and 13-1b.

Table 13-1a
Physical Requirements for Engine Oils
SAE Viscosity Grades for Engine Oils[a]—SAE J300 Mar93

SAE Viscosity Grade	Low-Temperature Viscosities		High-Temperature Viscosities		
	Cranking[b] (cP) Max at Temp °C	Pumping[c] (cP) Max with No Yield Stress at Temp °C	Kinematic[d] Min	(cSt) at 100°C Max	High Shear[e] (cP) at 150°C and 10[6] s[-1] Min
0W	3250 at -30	30,000 at -35	3.8	—	—
5W	3500 at -25	30,000 at -30	3.8	—	—
10W	3500 at -20	30,000 at -25	4.1	—	—
15W	3500 at -15	30,000 at -20	5.6	—	—
20W	4500 at -10	30,000 at -15	5.6	—	—
25W	6000 at -5	30,000 at -10	9.3	—	—
20	—	—	5.6	<9.3	2.6
30	—	—	9.3	<12.5	2.9
40	—	—	12.5	<16.3	2.9 (0W-40, 5W-40, 10W-40 grades)
40	—	—	12.5	<16.3	3.7 (15W-40, 20W-40, 25W-40, 40 grades)
50	—	—	16.3	<21.9	3.7
60	—	—	21.9	<26.1	3.7

[a]All values are critical specifications as defined by ASTM D 3244.
[b]ASTM D 5293.
[c]ASTM D 4684. The presence of any yield stress detectable by this method constitutes a failure regardless of viscosity.
[d]ASTM D 445.
[e]ASTM D 4683, ASTM D 4741, CEC-L-36-A-9O.

Table 13-1b
Physical Requirements for Engine Oils
Military Grades

	10W	30	40	5W-30	10W-30	15W-40
Cranking Viscosity[a] (cP) at Temp °C						
Min	3500 at -25	—	—	3250 at -30	3500 at -25	3500 at -20
Max	3500 at -20	—	—	3500 at -25	3500 at -20	3500 at -15
Pumping Viscosity[b] (cP) at Temp °C, Max	30,000 at -25	—	—	30,000 at -30	30,000 at -25	30,000 at -20
Viscosity[c] (cSt) at 100°C						
Min	5.6	9.3	12.5	9.3	9.3	12.5
Max	<7.4	<12.5	<16.3	<12.5	12.5	<16.3
Viscosity Index, Min	—	80	80	—	—	—
HTHS Viscosity (cP) Min	—	—	—	2.9	2.9	3.7
Pour Point (°C) Max	-30	-18	-15	-35	-30	-23
Stable Pour Point (°C) Max	-30	—	—	-35	-30	-263
Flash Point (°C) Min	205	220	225	200	205	215
Evaporative Loss[d] (%) Max	—	—	—	20	17	15
Phosphorus[d] (% Mass) Max	—	—	—	0.12	0.12	0.12

[a]ASTM D 2602 Modified.
[b]ASTM D 4684, allows no detectable yield stress.
[c]ASTM D 445.
[d]Not required for military specifications.

Multigrade oils are an outcome of the development of viscosity modifiers in the early 1950s. These are made by dissolving polymeric materials into low-viscosity, or thin, oils. These materials thicken oil more at high temperatures than at low temperatures. Hence, at low temperatures, oil flows easily due to its inherent low viscosity. But at high temperatures, its flow is resisted because of the thickening effect of the polymer. The invention of viscosity modifiers eliminated the need to use different viscosity oils for summer and winter operation and allowed the use of one oil during all seasons.

Gasoline Engine Oil (Service) Classifications

Until 1947, the performance of engine oils was defined solely by their viscosity grades, without consideration of engine design, its operating environments, and fuel type and quality. In 1947, API introduced three performance categories—regular, premium, and heavy-duty—based on severity of service. The regular oils were straight mineral oils with or without viscosity modifiers and corresponded to oils defined previously by the viscosity grades. These oils were for both gasoline and diesel engines operating under mild to moderate service conditions. The premium oils, designed for somewhat more severe operating conditions, generally contained oxidation and corrosion inhibitors and, in some cases, mild detergents. The heavy-duty oils possessed better oxidation and corrosion resistance and detergency than premium oils and were designed to withstand the most severe service.

Although it was a great improvement over that prior to 1947, this system did not distinguish oils based on fuel type. Therefore, in 1952, API introduced separate performance categories for gasoline and diesel engine oils. The gasoline engine oils were specified as ML, MM, and MS, and diesel engine oils as DG, DM, and DS. The first letters M and D stood for motor oil and diesel engine oil, respectively. The second letter represented the type of service. For motor oil, L stood for light, M for moderate, and S for severe service. For diesel engine oils, G stood for general use with no exceptionally severe requirements, M for moderately severe service, and S for exceptionally severe service.

The severity of service is a function of engine design and can vary across engine types, fuel characteristics, and operating conditions. MS type service was considered typical of gasoline and other spark-ignition engines that required protection against deposits, wear, and corrosion. This type of service is considered most severe because it includes both short trips (stop-and-go driving) and continuous driving at high speeds and at high loads (freeway or turnpike). Short trips promote oil screen and oil ring clogging, varnish deposits (especially on hydraulic valve lifters), and sludge formation in the crankcase, on rocker arm covers, and in oil filters. They also lead to corrosive wear or rusting of critical engine parts, such as cylinders, pistons, rings, and shafts, and to dilution of the oil by unburned fuel. Freeway operation promotes oxidation of lubricating oil which may lead to high-temperature varnish and sludge deposits, ring sticking and scuffing, and corrosion of some types of bearings.

To define oil quality for gasoline engines, a series of five sequence tests was adopted by API. Sequence I and II tests investigated low-temperature, medium-speed scuffing and wear and low-temperature deposit formation and rusting. Sequence III was a high-temperature oxidation test. A 1958 Oldsmobile engine was used for all three. Sequence IV investigated high-temperature, high-speed scuffing and wear using a 1958 DeSoto engine. Sequence V test assessed an oil's tendency to form insolubles and sludge and could use either a 1957 Lincoln engine or a CLR single-cylinder engine. Revisions to the API system occurred in 1955, 1960, and 1968. In 1969/1970, API collaborated with SAE and the American Society for Testing and Materials (ASTM) to install a new system of classifications that not only related to previous categories but also had the ability to provide new categories in the future. This classification had four categories, SA to SD, where S stands for service. Subsequently, other categories were added to parallel changes in lubrication environment as a result of changes in engine design, operating conditions, and testing procedures. The performance requirements of these categories are based on standard tests.

The API service symbol "Donut," established in 1983, communicates the engine oil's quality and performance to the general public. It helps in the selection of oils that meet manufacturers' recommendations for use in the intended application. The upper part of the symbol displays the API service category, the center part displays the SAE viscosity grade, and the bottom part displays the energy conserving feature, if applicable. Oils that achieve a minimum 1.5% reduction in fuel use relative to a reference oil are labeled as Energy Conserving I (EC-I), and those providing 2.7% reduction or over are labeled as EC-II. Only licensees are authorized to display this symbol. Of the categories listed in the API Classification System (SA to SH), only SH is active. The others are obsolete.

The International Lubricant Standardization and Approval Committee (ILSAC), a collaboration of the American Automobile Manufacturers Association and the Japan Automobile Manufacturers Association, has introduced its own performance categories, GF-1 and GF-2. These categories, issued in 1992 and 1996, use the API SH performance criteria but include a fuel economy Sequence VI test. The field testing is required only for technologies that are radically different from those in existence. The requirements of the obsolete categories are satisfied by API SH, ILSAC GF-1, and ILSAC GF-2, which are designed for the most severe operation. Gasoline engine oil categories along with the engine test requirements are given in Table 13-2. The time line is depicted in Figure 13-1. Note that Caterpillar 1H and 1H2 diesel engine tests, required for SD and SG, are not included in other categories, possibly because the combination categories, such as SF/CD and SH/CE, include a 1G or 1G2 diesel engine test.

Commercial (Heavy-Duty Diesel) Classifications

Compression ignition or diesel engines used before World War II were large, slow speed, and easy to lubricate. During and after the war, the need for fast-running engines led to the development of engines of compact but complex design. This placed a heavy demand on the lubricants. Excluding the SAE viscosity classification system, there were no established performance criteria and the end-user had to depend on the lubricant supplier's claims of the suitability of lubricants. As a consequence, OEMs started to introduce their own performance specifications. The Ingersoll-Rand Co. in 1931 was one of the first.

Table 13-2
Engine Oil Classification System for
Automotive Gasoline Engine Service
"S"—Service Oils

API Automotive Gasoline Engine Service Categories	Previous API Engine Service Categories	Related Industry Definitions	Engine Test Requirements
SA	ML	Straight mineral oil	None
SB	MM	Inhibited oil only	CRC L-4* or L-38 Sequence IV
SC	MS (1964)	1964 Models	CRC L-38 Sequence IIA* Sequence IIIA* Sequence IV* Sequence V* Caterpillar L-1* (1% sulfur fuel)
SD	MS (1968)	1968 Models	CRC L-38 Sequence IIB* Sequence IIIB* Sequence IV* Sequence VB* Falcon Rust* Caterpillar L-1* or 1H*
SE	None	1972 Models	CRC L-38 Sequence IIB* Sequence IIIC* or IIID* Sequence VC* or VD*

Table 13-2
(continued)

API Automotive Gasoline Engine Service Categories	Previous API Engine Service Categories	Related Industry Definitions	Engine Test Requirements
SF	None	1980 Models	CRC L-38 Sequence IID Sequence IIID* Sequence VD
SG	None	1989 Models	CRC L-38 Sequence IID Sequence IIIE Sequence VE Caterpillar 1H2*
SH	None	1994 Models	CRC L-38 Sequence IID Sequence IIIE Sequence VE

*The test is obsolete; engine parts, test fuel, or reference oils are no longer generally available, or the test is no longer monitored by the test developer or ASTM.

Military Specifications

Lubricant quality gained national importance during the war years, and a great deal of work was done to develop and test lubricants. In 1945, Caterpillar Tractor Co., which was involved in developing new diesel engines, collaborated with General Motors Corp. (GM) and, based on testing, listed a number of oils for their equipment. The U.S. military selected oils from this list for its use. In 1941, U.S. Army specification 2-104 was issued. The Coordinating Research Council (CRC), established in 1943, helped devise the first military specification, MIL-L-2-104A. It originally consisted of five tests, CRC L-1 to L-5. The L-1 test, of 480 h duration, evaluated lubricant performance in terms of ring sticking, wear, and

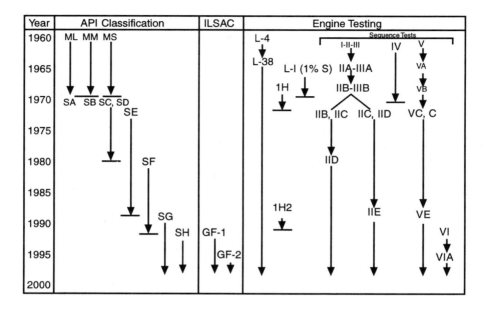

Fig. 13-1. History of performance classifications and engine requirements for gasoline engine lubricants.

deposits. The L-2 test, a 3-h 20-min long cycling test, evaluated accelerated running-in of the engine. Both tests used a single-cylinder Caterpillar Diesel engine. The L-3 test, designed to determine the stability of oil and its tendency to corrode copper-lead bearings, was 120 h long and used a four-cylinder diesel engine. The L-4 test, at 36 h, used a six-cylinder gasoline engine to determine oxidation characteristics of heavy-duty crankcase oils. The L-5 test, at 500 h, used a two-stroke, three- or four-cylinder diesel engine to determine oxidation, ring sticking, detergency, and silver bearing corrosion characteristics of the oil. After gaining experience, all tests except L-1 and L-4 were dropped. This led to military specification 2-104B in 1949. In 1950, the military specification MIL-O-2104 replaced specification 2-104B by taking into account fuel quality.

Caterpillar observed that MIL-O-2104 specified oils did not provide optimal performance in applications that involved lightly run and/or heavily loaded engines using high-sulfur fuels. High wear and poor performance were observed. The engine temperature was not high enough to evaporate water, which led to corrosion, and not low enough to prevent ring sticking.

The oils that performed well in the L-1 test with 1% sulfur fuel were called "Supplement 1" oils, and those that withstood both high sulfur and high temperatures as a consequence of supercharging were called "Supplement 2" oils. In 1954, the U.S. military adopted "Supplement 1" oils in its MIL-L-2104A specification. The L-4 test engine became unavailable; hence, the test was replaced with the CRC L-38 test. This test became part of the performance requirements of the MIL-L-45199A specification in 1961 and of the MIL-L-2104B specification in 1964. The MIL-L-45199A replaced the MIL-L-45199 specification of 1958 that was meant for oils for high-output diesel engines and was based on Caterpillar Series 3 lubricant requirements. MIL-L-2104B fulfilled the lubricant requirements of both supercharged and unsupercharged engines. The L-38 test, still in use, is a 40-h test that uses a Labeco CLR single-cylinder engine operating on leaded gasoline. Similar to its predecessor, it evaluates a lubricant's resistance to oxidation, as reflected by its viscosity change, and its sludge and varnish forming and bearing corroding tendencies. The specification MIL-L-2104C of 1970 resulted from the merging of MIL-L-45199A and MIL-L-2104B as it qualified oils for general heavy-duty use. This specification, in addition to standard tests (L-38 and Caterpillar 1G or 1G2) included gasoline sequence tests. In the D and E versions of MIL-L-2104, Mack and Detroit Diesel engine tests, and Caterpillar and Allison friction tests were included. The inclusion of the friction tests makes these oils suitable for use in transmissions also. The MIL-L-2104F, introduced in 1991, is the most recent of the military specifications.

The Caterpillar engine test similarly evolved over the years. The L-1 engine test using a 3400 cm^3 Caterpillar engine was established in 1943. The test, 480 h long, was run at an engine speed of 1000 rpm. It evaluated a lubricant's ability to control lacquer, deposit formation, and wear. In 1948, it was replaced by the 1D test that used a supercharged engine. In addition, the speed of the engine was increased to 1200 rpm. The 1G test, established in 1958, used a downsized 2200 cm^3 engine operating at an even faster speed of 1800 rpm. In 1956, Caterpillar introduced Superior Lubricants (Series 3) to eliminate field problems of supercharged engines. These lubricants were required to pass both Caterpillar 1G and 1D engine tests. Test procedure change led to the transition from 1G to 1G2 around

1976. The test duration for all these tests is 480 h. The 1K and 1N tests, the most recent of the Caterpillar single-cylinder engine tests, are of 252 h duration and use a low-emission piston design.

API Classification System

This classification for commercial or C oils for diesel engines resulted from the collaboration of API/ASTM/SAE in 1969/1970. Originally, four categories, CA to CD, were introduced. Over the years, new categories were added to meet the changing lubrication needs of engines due to changes in design and/or operating conditions. At present, the categories range from CA to CG-4. Of these, all except CF, CF-2, CF-4, and CG-4 are now obsolete.

OEM Specifications

The commercial oils available in 1972 did not have adequate performance in Mack trucks. Therefore, Mack issued EO-H specification to qualify oils for its equipment. As Mack modified its engines, the specification evolved into EO-J and EO-K/2 over the 1972 to 1984 period. The EO-H qualified oils were to pass Mack T-1, the EO-J oils were to pass Mack T-5, the EO-K oils were to pass Mack T-6, and the EO-K/2 qualified oils were to pass both Mack T-6 and T-7 engine tests. This requirement for improved oil quality paralleled Mack's efforts to improve oil consumption, fuel economy, and emissions through engine modifications.

Caterpillar tests saw a similar progression. Its tests evolved from L-1 to 1N via 1D, 1G, 1G2, 1H, 1H2, 1K, and 1M-PC. Of these, 1K and 1N use low-emission piston designs and relate to OEM efforts to control undesirable emissions. The 1M-PC test, a 120-h version of the 480-h long 1G2 test, could become a part of upcoming Service categories. Cummins NTC-400 and the Detroit Diesel Corporation have added their own specifications for similar reasons.

The evolution of diesel engine oil specifications is depicted in Figure 13-2. The API categories are given in Table 13-3, along with the required engine tests. The engine test requirements for all diesel specifications, including those of the U.S. military and Mack, are given in Figure 13-3. The history of diesel engine testing is shown in Figure 13-4. In addition to pure categories, certain factory-fill and military lubricants previously had combination specifications, such as SH/CE. The oils meeting these requirements could be used in both gasoline and diesel engines.

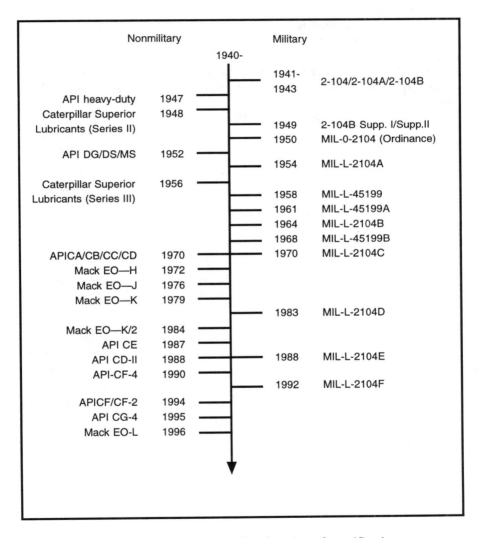

Fig. 13-2. History of American diesel engine oil specifications.

Table 13-3
Engine Oil Classification System for Commercial Diesel
Engine Service
"C"— Service Oils (Fleets, Contractors, Farmers, Etc.)

API Automotive Gasoline Engine Service Categories	Previous API Engine Service Categories	Related Industry Definitions	Engine Test Requirements
CA	DG	MIL-L-2104A	CRC L-38 Caterpillar L-1* (0.4% sulfur fuel)
CB	DM	Inhibited oil only	CRC L-38 Caterpillar L-1* (0.4% sulfur fuel)
CC	DM	MIL-L-2104A Supplement 1	CRC-L-38 Sequence IID Caterpillar 1H2*
CD	DS	MIL-L-2104B MIL-L-46152B	CRC L-38 Caterpillar 1G2
CD-II	None	MIL-L-45199B, Series 3 MIL-L-2104C/D/E	CRC L-38 Caterpillar 1G2 Detroit Diesel 6V53T
CE	None	None	CRC L-38 Caterpillar 1G2 Cummins NTC-400 Mack T-6 Mack T-7
CF-4	None	None	CRC L-38 Cummins NTC-400 Mack T-6 Mack T-7 Caterpillar 1K

**Table 13-3
(continued)**

API Automotive Gasoline Engine Service Categories	Previous API Engine Service Categories	Related Industry Definitions	Engine Test Requirements
CF-2	None	None	CRC L-38 Caterpillar 1M-PC Detroit Diesel 6V92TA
CF	None	None	CRC L-38 Caterpillar 1M-PC
CG-4	None	None	CRC L-38 Sequence IIIE GM 6.2L Mack T-8 Caterpillar 1N

* The test is obsolete; engine parts, test fuel, or reference oils are no longer generally available, or the test is no longer monitored by the test developer or ASTM.

Gear Oils

All gear systems use one or more of the four basic types of gears. These are spur and helical, bevel, hypoid, and worm. Proper lubrication is critical to the optimum life of a gear system. This was realized early, as indicated by the 1909 recommendation for the rear axle lubricants to be of reasonable viscosity. Prior to the introduction of the hypoid gear by Gleason Works in 1925, passenger car final drives employed spiral bevel gears which, under normal use, could be lubricated with straight mineral oil. Hypoid gears gained popularity in the late 1920s because they allowed automobile designs of a lower center of gravity, quieter operation, and reduced size and weight of the differential assembly. Because these gears have sliding action in addition to rolling action, they have prolonged contact and hence require better lubricants than needed for spiral bevel gears. Lubricants containing a lead soap-active sulfur additive combination,

	Caterpillar Series 3	MIL-L- 45199	API CD	API CE	API CD-II	API CF-4	API CF	API CF-2	API CG-4
YEAR	1956	1958	1970	1987	1988	1991	1994	1994	1994
Caterpillar 1D	■	■	■						
Caterpillar 1G	■	■	■						
Caterpillar 1G2				■	■				
CRC L-38		■	■	■	■	■	■	■	■
Sequence IIB									
Sequence IIC									
Sequence IID									
Sequence IIC									
Sequence IID									
Sequence IIE									■
Sequence VC									
Sequence VD									
Sequence VE									
MIL-L-2104C									
Mack T-1									
Mack T-5									
Mack T-6				■		■	■		
Mack T-7				■		■	■		
Mack T-8									■
DDC 6V53T					■				
DDC 6V92TA								■	
NTC 400				■		■	■		
Caterpillar 1K						■	■		
Caterpillar 1M-PC								■	
Caterpillar 1N									■
GM 6.2L									■
Caterpillar TO-2									
Caterpillar TO-4									
Allison C-3									
Allison C-4									

Fig. 13-3. Heavy-duty diesel engine test requirements.

218

	MIL-L-2104C						MIL-L-2104D	MIL-L-2104E	MIL-L-2104F	MACK				
YEAR	1970	1972	1974	1976	1978	1980	1983	1988	1991	1970	1972	1976	1979	1984
Caterpillar 1D	●	●	●	●	●									
Caterpillar 1G	●	●	●											
Caterpillar 1G2				●	●	●	●	●						
CRC L-38	●	●	●	●	●	●	●	●	●					
Sequence IIB	●													
Sequence IIC		●	●	●										
Sequence IID					●	●	●	●						
Sequence IIC			●	●										
Sequence IID					●	●	●							
Sequence IIE								●	●					
Sequence VC	●	●	●	●	●									
Sequence VD						●	●							
Sequence VE								●						
MIL-L-2104C										●				
Mack T-1											●			
Mack T-5												●		
Mack T-6													●	●
Mack T-7									●					●
Mack T-8														
DDC 6V53T							●	●	●					
DDC 6V92TA														
NTC 400														
Caterpillar 1K								●						
Caterpillar 1M-PC														
Caterpillar 1N														
GM 6.2L														
Caterpillar TO-2							●	●						
Caterpillar TO-4									●					
Allison C-3							●							
Allison C-4									●					

Fig. 13-3. (continued)

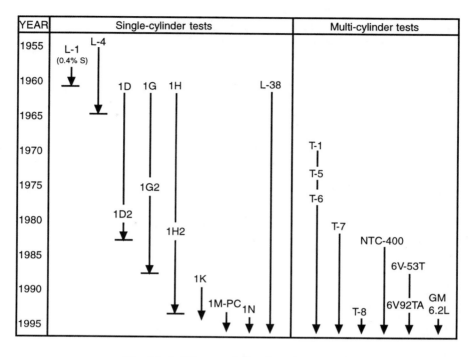

Fig. 13-4. History of diesel engine testing.

known as an extreme-pressure, or EP, additive system were found to be effective. When GM in its 1937 models introduced hypoid axles, it tested and recommended the use of a number of such lubricants. The EP agents are believed to react with metal surfaces at high temperatures to form protective chemical films.

Classification Systems

Gear oils are formulated primarily to provide extreme pressure protection for gears and axles to prevent fatigue, scoring, and wear damage under boundary lubrication conditions. Automotive gears perform a number of diverse functions. Because of differences in design, each gear type places different demands on the lubricant. Automotive gear oils are classified in a manner similar to that of engine oils, i.e., by SAE viscosity grades, API service designations, U.S. military specifications, and OEM performance requirements.

Table 13-4
Physical Requirements for Gear Lubricants Intended for Axle and Manual Transmission Applications

Property	SAE Viscosity Grade						MIL-L-2105D		
	75W	80W	85W	90	140	250	75W	80W-90	85W-140
Viscosity (cSt) at 100°C									
Min	4.1	7.0	11.0	13.5	24.0	41.0	4.1	13.5	24.0
Max				<24.0	<41.0			<24.0	>41.0
Max Temp (°C) for 150,000 cP Viscosity	-40	-26	-12	—	—	—	-40	-26	-12
Channel Point (°C) Max	—	—	—	—	—	—	-45	-35	-20
Flash Point (°C) Min	—	—	—	—	—	—	150	165	180

Viscosity has an impact on the load-carrying capacity and hence wear of the equipment. The load-carrying capacity is the maximum load that a sliding or rolling system can support without failure. In the early 1920s, SAE introduced the first viscosity-based classification system for transmission and rear axle oils. It included grades 80, 90, 110, and 160, which were based on SUS viscosities of oils at 98°C (210°F). Later, grades 110 and 160 were replaced by 140. Over the years, the system evolved to its present status, as provided in Table 13-4.

U.S. Military Specifications

In the 1940s, API used designations regular, worm, mild, and multipurpose to describe gear lubricants. Because multipurpose gear lubricants did not provide satisfactory performance, the U.S. Army took the initiative in the 1940s to develop an effective gear lubricant. It used federal specification VV-L-761 for "Lubricants, Enclosed Gear, Hypoid Gear and Other Types" as the basis for its 2-105A specification. World War II helped identify deficiencies in gear lubricants when used under low-speed high-torque service. This led to the development of more effective multipurpose lubricants.

At the request of the U.S. military, CRC introduced two procedures to evaluate the EP properties of gear lubricants. The L-19-645 procedure evaluated lubricants under high-speed operation (to simulate highway driving), and the L-20-545 procedure evaluated a lubricant's effectiveness under high-torque, low-speed operation (to simulate mountain operation with high load). The gear distress (surface damage) was described as rippling, ridging, pitting, scoring, and spalling. These tests became part of the Army specification 2-105B of 1946. It was renamed as MIL-L-2105 in April 1950. The original L-19 test procedure, devised for a road test, was modified to run the test on a dynamometer. The L-19 test used a Chevrolet passenger car with a repeated cycling speed of 60 and 80 mph. The L-20 test used a U.S. military 3/4-ton truck which was run at a ring-gear speed of 62 rpm at a 6000 lb-in torque for 20 min, followed by 30 h at 32,311 lb-in torque.

In the early 1950s, the heavy-duty military trucks started to show gear tooth distress (ridging), suggesting that MIL-L-2105 qualified lubricants did not perform well in the field. After extensive testing, the L-20 test was replaced with the CRC L-37 test that included a higher torque and an initial high-speed sequence to simulate the field performance. At the same time, passenger car manufacturers requested a more severe test than L-19 because of concern over an imbalance between scoring and heavy-duty protection. CRC used the passenger car manufacturers' data on scoring to establish the L-42 high-speed test. These two tests became part of the 1959 specification MIL-L-2105A. The MIL-L-2105B specification of 1962 included a thermal oxidation stability test in addition to the L-37 and L-42 axle tests. This test measured a lubricant's viscosity increase and tendency to form pentane and benzene insolubles at high temperatures. Concern over marginal rust and oxidation performance of MIL-L-2105B qualified oils and the emergence of multigrade oils in 1976 led to the next upgrade, MIL-L-2105C. It included oils of viscosity grades 75W, 80W-90, and 85W-140 and a more severe moisture corrosion test (L-33 test). This specification superseded both MIL-L-2105B and the 1957 specification

MIL-L-10324A that was designed for lubricant use at low temperatures. In 1987, MIL-L-2105C was replaced by MIL-L-2105D that has the same test requirements as the C version, except that the thermal oxidation stability test is CRC L-60, which uses toluene insolubles instead of benzene insolubles. The specification MIL-Prf-2105E, issued in 1995, combines the GL-5 requirements of MIL-L-2105D and thermal oxidation stability, anti-wear, and seal compatibility requirements of the newly released API specification MT-I for nonsynchronized manual transmissions used in buses and heavy-duty trucks.

API Service Categories

The API service categories for automotive manual transmissions and axles were first published in 1966. Of the original six, GL-1 to GL-6, only five exist today. GL-6 is obsolete. The GL-4 and GL-5 categories, which correspond to U.S. military specifications MIL-L-2105 and MIL-L-2105D, define oils for service-fill only. Factory-fill oils are defined by major car and truck manufacturers. Such oils have performance characteristics that are critical to the satisfactory operation of a particular drivetrain and may include break-in, bearing preload, and limited slip durability. Mack GO-G for heavy-duty truck lubricants, introduced in 1975, is one such specification. This evolved into GO-H in 1990. The specifications GO-G/S and GO-H/S for synthetic fluids came out in 1985 and 1990, respectively.

Gear oils are tested according to the methods established by the American Coordinating Research Council (CRC) in specified axles. Table 13-5 presents API service designations, and Table 13-6 presents the affiliated bench and axle tests. API MT-1, introduced in 1995, defines lubricants for nonsynchronized manual transmissions and emphasizes thermal stability, wear control, and seal compatibility.

Transmission Fluids

The key functions of these fluids are lubrication, cooling, and to act as a hydraulic medium to transmit power. There is no API classification for these fluids. OEMs, such as GM and Ford Motor Co., establish the performance requirements for automatic transmission fluids (ATFs). These deal with frictional consistency (durability), frictional compatibility with the transmission's components (such as clutches, bands, and synchronizers), oxidation, and wear. Manual transmissions use a variety of fluids including ATFs, engine oils (5W-30), some gear lubricants, and specialty fluids. In addition to frictional properties, certain OEMs require the transmission fluids for their equipment to have improved shear stability, low-temperature fluidity, and other specific characteristics.

Table 13-5
API Gear Oil Service Designations

API Classification	Type	Applications
GL-1	Straight mineral oil	Truck manual transmissions
GL-2	Usually contains fatty materials	Worm gear drives, industrial gear oils
GL-3	Contains mild EP additives	Manual transmissions and spiral bevel final drives
GL-4	Equivalent to obsolete MIL-L-2105; usually satisfied with 50% GL-5 additive level	Manual transmissions, and spiral bevel and hypoid gears in moderate service
GL-5	Virtually equivalent to present MIL-L-2105D; primary field service recommendation of most passenger car and truck builders worldwide.	Moderate and severe service in hypoid and other types of gears. May also be used in manual transmissions.
GL-6	Obsolete	Severe service involving high-offset hypoid gears
MT-1	Contains thermal stability and EP additives	Nonsynchronized manual transmissions in heavy-duty service

Table 13-6
API GL-5 and MIL-L-2105D Requirements

Test	Description	Characteristics Measured
CRC L-33	Gear test using differential assembly	Resistance to corrosion in the presence of moisture
CRC L-37	Gear test using complete axle assembly	Resistance to gear distress under low-speed, high-torque conditions
CRC L-42	Gear test using complete axle assembly	Resistance to gear distress (scoring) under high-speed, shock-load conditions
CRC L-60	Bench test using spur gears	Thermal oxidation stability
ASTM D 892	Bench test	Foaming tendencies
ASTM D 130	Bench test	Stability in the presence of copper and copper alloys

Passenger car automatic transmissions were introduced in the United States in 1939. Although special lubricant needs were realized as early as 1937, in the absence of such lubricants, widely available engine oils and mineral oils were used. This practice continued until 1949 when GM introduced its first specification for these fluids.

GM Specifications
In 1937, Oldsmobile Division of GM installed the first automatic transmission, which was originally lubricated by engine oils. These oils proved unsatisfactory and were replaced by a specially formulated lubricant both for factory-fill and service-fill. The first official specification, "Automatic Transmission Fluid, Type A," was introduced by GM in 1949. The field performance of Type A fluids was inadequate, and the specification was upgraded to Type A, Suffix A. This specification included a powerglide test to assess the lubricant's resistance to oxidation. The deficiencies in

Type A, Suffix A lubricants became apparent in the ensuing years and, in 1967, GM introduced the Dexron specification. The requirements of Dexron fluids were similar but more stringent than Type A, Suffix A fluids. Also included were a low-energy, cycling friction retention and oxidation test and a high-energy, transmission cycling, friction retention and durability test. The first test dealt with maintaining shift time retention under severe service conditions; the second test dealt with transmission clutch durability under heavy-load operations.

Transmission design changes and increased severity of service required improved fluid durability. This led to the introduction of the Dexron-II specification in 1973, which included four new tests: Turbo Hydramatic Oxidation Test (THOT); Turbo Hydramatic Transmission Recycling Test (THCT); High Energy, Friction Characteristics and Durability Test (HEFCAD); and a wear test. The Dexron-III upgrade of 1994 was in response to transmission design and operating environment changes to improve fuel economy and emissions, all of which lead to higher operating temperatures. Dexron-III qualified fluids, therefore, must have good low-temperature properties and better oxidative stability and frictional durability than Dexron-II fluids.

Ford Specifications

Prior to 1959, all major OEMs used GM Type A specified oils, although they were not suitable for all equipment. Therefore, in 1959 Ford established its own specification M2C33-A/B for its factory-fill which was similar to the GM Type A, Suffix A. The first suffix letter in the Ford specification is for the undyed fluid; the second suffix letter is for the red-dyed fluid. In 1958, Ford set up a goal to introduce "fill-for-life" fluids for its transmissions. Neither Type A, Suffix A nor M2C33-A was suitable because of inconsistent performance and because automatic transmissions were experiencing increasing loads during use. Ford introduced a new specification, M2C33-C/D, in 1961. The M2C33-C/D qualified fluids had superior oxidation, improved wear, and higher static capacity than those in existence. In 1967, Ford added a friction requirement to the M2C33-D specification, which resulted in the M2C33-F, or simply Type F, specification. In 1982, specification M2C166-H replaced Type F to qualify oils for use in the new Ford C-5 transmission. The Mercon specification of 1987 was to select fluids for the 4L60 transmissions. These fluids are required

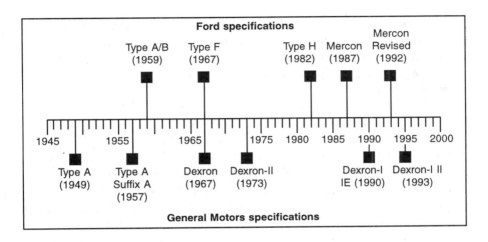

Fig. 13-5. ATF service-fill specification time line.

to have better thermal stability and improved performance in the plate clutch and cycling tests. This specification has undergone several revisions, the most recent of which was in 1994. The time line for the introduction of the ATF specifications is given in Figure 13-5.

OEM Requirements

The Allison Transmission Division of GM and Caterpillar specify fluids for their own equipment, normally used for off-highway, heavy-duty applications. They consider automatic transmission fluids as unsuitable for the more severe operating environments of such vehicles. Allison C-4 and Caterpillar TO-4 are the most recently issued of such specifications.

Greases

The use of animal fat in combination with lime can be traced to ancient times. The word "grease" is derived from the Latin word "crassus" for fat. Lubricating grease is defined in ASTM D 288 as "a solid-to-semifluid product of dispersion of a thickening agent in a liquid lubricant. Other ingredients imparting special properties may be added." Although a variety of agents can disperse in oil to yield grease-like dispersions, only those that form dispersions with lubricating properties are useful. In addition, grease contains additives that impart desirable properties, such as EP, demulsibility (resistance to water), etc. The lubrication function is carried out by the small amount of oil that is released during equipment operation. Because of their semisolid nature, greases are used when fluid lubricants are inefficient, the need for lubrication is infrequent, and/or the lubricant is required to maintain its original position in a mechanism.

Lime or calcium soap (calcium carboxylate) greases can be traced to the 1880s. They were among the first to be produced and marketed in volume. Their manufacture involved cold mixing of lime with rosin oil (acid) in mineral oil. The calcium soap that formed imparted a greasy consistency to the mixture. These were replaced by preformed calcium-soap greases that were smoother, more water-resistant, and had superior lubricating properties. Because these greases require 1–2% water to stabilize them, they are not appropriate for applications involving temperatures over 100°C (212°F). The accompanying loss of water leads to their deterioration. This eliminates their use in modern automotive equipment that operates at high speeds and high operating temperatures.

Sodium soap greases, developed in the 1930s, did not suffer from this problem and had better resistance to high temperatures. They also were more stable to mechanical handling. Although these greases were used to lubricate automotive wheel bearings, they were replaced by better water- and heat-resistant multipurpose greases. These include complex calcium soap greases (invented in 1940), lithium soap greases (invented in 1942/1943), and barium greases (invented in the 1950s). Lithium and barium soap greases retain their semisolid characteristics at much higher temperatures than their sodium and calcium analogs. These greases usually contain sodium nitrite to inhibit corrosion and oxidation. A lithium complex grease with even better high-temperature properties has become available since 1962.

Most sodium and calcium greases are made from animal fats and/or fatty acids. Lithium soap greases are made from hydrogenated castor oil or the corresponding 12-hydroxystearic acid. Complex soap greases are made from a mixture of fatty and nonfatty acids and have better water and heat resistance than simple soap greases. Lithium soap greases account for more than 50% of today's total grease production.

In addition to soaps, polymeric and inorganic thickeners can be used. These include polymers such as poly(urea), poly(ethylene), poly(carbohy-drate), and poly(tetrafluoro-ethylene) or Teflon, and inorganics such as modified clay (bentonite), graphite, carbon black, molybdenum disulfide, and various metal oxides. Modified clay and poly(urea) based thickeners are gaining popularity because they result in highly water- and heat-resistant greases.

These lubricants are defined according to the National Lubricating Grease Institute (NLGI) Service Classification System, first implemented in 1991. It describes chassis and wheel bearing greases according to ASTM D 4950 of 1989. LA and LB are classes for chassis greases; GA, GB, and GC are classes for wheel bearing greases. LA is for mild duty, frequent relubrication service; LB is for infrequent relubrication, high loads, water exposure type service. GA is for mild duty, GB is for moderate duty, and GC is for severe duty service. Prior to this classification, the SAE recom-mended practice, published in the SAE information report J310, was used for this purpose. The report, first introduced in 1951, had several revisions, the most recent of which occurred in 1993.

Materials: Key to 100 Years of Automotive Progress

Gary Bragg
Caterpillar Inc.

"A little business at Carlisle, a little curiosity concerning Gettysburg, a little visit at Hagerstown, a general longing for an outing, and a desire to give the three-wheeler a long drive over all kinds of roads, prompted a trip toward the Shenandoah Valley...proved too much for one of our springs, resulting in a broken leaf...

"Here we stopped for dinner, while the carriage was receiving the new spring ordered to meet us at this point. This new spring proved to be but little better than the old one, so we decided to hunt up a blacksmith and have the low spring stiffened by inserting another plate. The matter was made worse rather than better, because the temper was now out of the original spring and the new leaf had none...Although we had never experienced such a sinking sensation before this was diagnosed as a broken axle, which investigation with a lighted match proved...About 10 miles out of Carlisle, a loud report indicated a burst tire. This was repaired by a spare inner tube and stout wrapping of string and tire tape...At Harrisburg, we bought a supply of string and tire tape with which to mend our damaged tire better...When well out of town, we stopped at a quiet place, cut off the original wrappings—rewrapped, spending about an hour...We had completed about 100 miles that day and had not driven steadily, either." The trip was voted a decided success—Charles Duryea, *The Horseless Age*, 1903.

That was a typical trip in the primitive early days of the American automotive industry. Most of Duryea's troubles related to materials failures. The relationship between product design, materials design, and manufacturing can always be questioned. The word "materials" means different things to different people.

This Duryea Trap is similar to the one Charles Duryea drove on his weekend trip.

To a designer, "materials" are sometimes a necessary afterthought that limits the design. To others, they are the objects of all their attentions as they try with difficulty to form them into useful shapes. A few people spend their lives understanding material flaws, stresses at the crack tip, and cyclic strain energy. To many, the word "materials" conjures the sensation of texture, feel, richness, and refinement.

As J.H. Dickenson stated very well in 1910: "Heated arguments can be imagined as arising, ages ago, over the ruins of a collapsed mud hut in Asia Minor, which would form an almost precise parallel to some present-day controversies over such matters as, for example, the failure of a drop stamped part of a motor car."

It would probably solve the dilemma of what "materials" are if an automotive design team from today were somehow transported by time machine back to around 1896, the beginning of the American auto industry. Then, using the materials and manufacturing available in 1896, and today's knowledge, could the team design a credible car that approximates a contemporary car? Of course, this will never happen, but speculation on this phenomenon would give clues to the primary character of the words "automotive materials."

In both popular automotive literature and technical papers, materials are given minimum time in the spotlight. Only in specialized papers are materials dominant.

To accomplish this story, the development of the American automobile will be divided into somewhat arbitrary eras describing automobile design—not materials' development, for automotive materials development can never be more than the design of the automobile as a whole.

To produce a definitive, all-encompassing history of automotive materials development is an impossible task for a brief article. However, we owe it to the metallurgists, chemists, and foundrymen who, without much recognition, supported and many times provided the means to bring the automobile to its present state of safety, convenience, and comfort.

Although this story covers primarily the American automotive industry, it is impossible to ignore the changes caused by the simultaneous development of the automobile in other parts of the world.

The Primitive Years: 1896 to 1906

As seen through the eyes of the would-be American automobile inventor, 1896 America brimmed with industrial resources and materials: steam power, railroads, ships, machine tools, farm implements, and a growing electrical power grid. Unfortunately, most equipment was heavy and clumsy. Castings were massive so that the unknown stresses would be low. Irons and steels of high quality could be obtained, but the production of this high quality was more an art than a science.

The only vehicles that permitted unrestrained personal travel were pulled by horses, and the materials for those vehicles had been around for hundreds of years. Wrought steels, wood, canvas, and leather composed the bulk of any carriage or wagon. These materials allowed the lightweight design required for any personal vehicle, and they were plentiful.

For quietness, laminated and glued leather gears were utilized. If there was a concern for weather protection, cellulose nitrate could be used in canvas curtains as they had been for buggies since 1869.

Casting technology was developed to the point that "green sand" castings using molding boxes had been in use for a hundred years and cast steel was becoming commercially available. Die castings were being used in cash registers and phonographs as well as in typesetting equipment. Although the early inventors relied on materials available around the neighborhood machine shop, smithy, and foundry, the impact of the new industry was evident even in these early years.

Franklin started using die cast bearings for connecting rods in 1904. From that year, the automotive industry has been the single biggest user of die castings.

As the comments of Charles Duryea suggest, pneumatic tires were available at the start of the automobile industry. The development of the tire of today started in the 1800s with the bicycle craze. It could be debated which spawned more automobile start-ups—buggy or bicycle shops.

While state-of-the-art materials processes and know-how were available in these primitive years in the firearm, textile, sewing machine, and machine tool industries, the early auto pioneers were more or less limited to local buggy shops, machine shops, and foundries.

However, an explosion of events was soon to change swiftly the image and substance of the new industry. One significant event happened on January 15, 1906, at New York's New Grand Hotel—the first annual meeting of the Society of Automobile Engineers (SAE). Two of the first papers were *Materials for Motor Cars,* by Thomas J. Fay, a consulting engineer from Brooklyn, New York, and *Ball Bearings,* by Henry Hess of Hess-Bright Mfg. Co. of Philadelphia, Pennsylvania.

The first steps in standardization were also taken early in the life of the industry. The National Association of Automobile Manufacturers initiated the first few standards for the industry in the form of reports on physical tests of tires, screws, steel tubing, iron, and steel.

The Brass Age: 1907 to 1914

By 1907, the automobile was taken seriously as a replacement for the horse. People such as Packard, Leland, and Pierce had already established reputations as quality automobile builders. Better financed, larger automotive operations were starting up that could tap into the state-of-the-art technologies of that day.

In 1910, SAE organized the General Standards committee to take over the pioneering standards work of the National Association of Automobile Manufacturers. By January 1911, SAE specification No. 1 was issued. This first specification covered 0.50% carbon steel, which is now known as SAE 1050.

At the 1911 SAE summer meeting in the Algonquin Hotel at Dayton, Ohio, the Dayton Chamber of Commerce presented every SAE member with a souvenir binder for all the loose "standards." Thus, the first *SAE Handbook* was born.

General Motors began its first metallurgical laboratory in 1911. Some people thought something could be gained by looking through a microscope at a piece of steel. The piles of failed parts kept growing. The average motorist could either count on fatigued springs and crankshafts or massively over-designed, clumsy components.

Brass, bronze, and aluminum, prized today by the collector, were used because they were easily cast and worked—not because of their beauty. Often, these handsome materials were painted.

Ford began producing vanadium steel in large quantities for use in the 1907 Model T. With Harold Wills' help, Ford was able to build a lighter but stronger automobile with the new alloy steel.

Ford found that the new steel was three times stronger, but more machineable, than its standard steels. This new material was critical to Ford's vision of a low-cost, lightweight automobile for the masses. Although the use of vanadium alloy was pioneered in Europe, it took Wills and Ford to manufacture it in large economical heats.

In 1907, disaster struck the zinc die-casting business. Die-cast parts all over the country were swelling out of shape and crumbling. Lead and tin impurities in the alloy were causing electrolytic action that would eventually cause the part to lose functionality. It was not until the 1920s that zinc die casting became reliable. These old die castings turned out to be the bane of present-day restorers.

The need for close-tolerance, low-cost, durable die castings in the auto industry was obvious. Aluminum die castings were ideal in many ways, but molten aluminum would dissolve the iron in the die-caster pump, damaging the machine.

By 1910, primitive and dangerous aluminum die casters were turning out 10,000 shots from a single die—practical, but far from optimum. The Great War caused a huge demand for die castings for everything from machine-gun gears to gas masks. This spurred the development of improved die casting techniques in the 1920s and 1930s. Cold-chamber machines came into favor, in which the metal for each shot is ladled into the relatively cold chamber.

An automobile is no better than its tires. The explosion of the automotive industry spawned the rubber plantation. In 1907, almost all the rubber for tires came from the wild and lacked even the addition of carbon black for strength. By 1914, 60% of the rubber was coming from plantations, and carbon black was being introduced into tires by Diamond Rubber and B.F. Goodrich.

In 1912, Edward Budd started a small body-building company to supply the growing auto industry with all-steel bodies. People such as Budd saw real opportunity in the developing automobile industry. Indeed, auto manufacture increased 60% over 1911, reaching 350,000 units. Budd's first two customers were Garford Motors and Oakland. To build all-steel bodies, newly developed acetylene welding equipment was imported from France.

At the same time, electric processes were being developed. Budd welded large sheets of steel by applying current to two large sheets butted together in large electrode jaws. As the sheets touched and the mating surfaces held under pressure, a flash of light occurred and a strong uniform joint resulted.

During this era, the value of standardization became accepted. Standardization of seamless tubing brought the number of sizes from 1600 to 221. One manufacturer had previously used 80 sizes on one vehicle.

Then, as it is now, some people were opposed to standards on the grounds that standards regimentation would hinder creative ideas. However, most engineers viewed the new standards as a means to avoid mindless repetition and unnecessary expense.

Everyman's Car: 1915 to 1925

By 1915, forward-thinking people were predicting the day that cars and trucks would be strong competitors to the railroad industry. Motor cars were becoming more powerful all the time, and the aeroplane proved itself in the Great War.

By 1924, the United States had 50,000 km (31,000 mi) of concrete roads. In 1925, a standardized highway numbering system was adopted. America was on the move by automobile.

The Automobile: A Century of Progress

These advancements demanded an integration of materials development and design. A broken crankshaft or axle could not be tolerated in a vehicle that was then being depended on for daily transportation.

Wohler, in 1871, had identified the importance of stress concentrators in his famous rotating-beam railway axle tests. Basquin, as early as 1910, had related fatigue strength properties to monotonic and cyclic stress/strain properties.

By 1915, designers and metallurgists were openly talking about the dangers of stress concentrators and weighing the merits of alloying steels for toughness—where the extra cost would pay off and where it would not. The discussions became rather sophisticated and advanced, with words such as slip planes and flaws that were very localized in nature.

However, in 1915, engineers and metallurgists were still trying to discover if fatigue cracks could be healed. In the case of hardened parts, they did not understand the important role residual stresses play in the relationship of case and core.

In this era, designers were concerned with minimizing the weight of the automobile to provide economy and durability. As much as 23 million kg (25,346 tons) of aluminum were used in a single year for crankcases, transmission housings, and body components.

A British engineer, Pomeroy, designed and built a car, based on the Pierce Arrow, that made full use of aluminum. The resulting vehicle weighed two-thirds as much as a comparable car. In the United States, Franklin and Dodge were using aluminum connecting rods. In Europe, Hispano-Suiza had aluminum blocks.

The first truly synthetic thermosetting plastic came into use during this time. In 1909, Dr. Leo Baekeland had developed plastic from a combination of pheno-formaldehyde and ammonia. Named Bakelite in his honor, this material became the ubiquitous distributor and rotor.

Iron-casting quality was being improved throughout the 1920s. More powerful and refined multicylinder autos demanded good castings. Sand testing and control were established, and in 1923, synthetic sand mixtures were introduced.

A brushed-aluminum body trimmed with black walnut was featured on this 1922 Velie.

In 1924, DuPont developed a pyroxylin lacquer called Duco for the Oakland. This new paint reduced painting time from days to minutes and produced a finish that was far superior to the hand-brushed varnish then in use—a holdover from carriage days.

Of equal importance, the new paint brought a great array of colors and the start of the reign of style over the automobile. Colors such as Liqueur Yellow, Alhambra Tan, Badminton Green, and Star Thistle Blue soon would bring visions of dazzling splendor to an otherwise dull world.

The Classic Car: 1925 to 1939

This era began with the universal, but now dated, Model T continuing to hold sway, and ended with the autos of the late 1930s that could be considered as reasonable transportation even today. Simplicity, dependability, and short economic life characterized the cars of the late 1920s.

Researchers had finally decided that martensitic steel was a solid solution of carbon in iron. X-rays had determined that the lattice of martensite was body-centered tetragonal. Heat treating of steels was becoming a science rather than an art.

With more and more concrete highways allowing safe 100-km/h (62-mph) speeds for hours at a time, higher engine dynamic loads had to be addressed. In 1934, the first autos using Clevite's continuously cast, copper-lead bearings were hitting the roads. These new bearings demanded better tolerances and cleanliness. No longer could thick-wall babbitt bearings be expected to run in and "eat" debris.

Aluminum pistons very successfully replaced cast iron in most auto engines. Aluminum cylinder heads were used less successfully. While engine performance increased, the car owner paid a price in durability.

However, as the multicylinder race started in the early 1930s, weight was found to be more a selling feature than light, efficient cars. Aluminum housings and body panels were dropped in favor of cast iron and steel.

This 1932 Packard engine, with its aluminum crankcase, pan, and transmission housing, was a leftover from 1920s design.

However, there were exceptions. Howard Marmon employed aluminum for the cylinder heads, block, and inlet manifolds of his wet-linered V16 engine.

In 1936, an event occurred in Germany that would, decades later, give the U.S. auto industry tremendous anxiety. That event was the design of the Volkswagen with 18 kg (40 lb) of magnesium castings. With magnesium used for the crankcase and transmission housing plus other smaller parts, 50 kg (110 lb) were taken from the car's mass. This amounted to 7% of the car's total mass.

The gradually increasing use of rubber in the automobile was also significant. When Chrysler placed its four-cylinder engine into optimally located rubber mounts and called it Floating Power, the path was opened to provide vehicle occupants with vibrationless comfort.

In February 1935, experimenters at DuPont first synthesized nylon in an effort to mass produce a substitute for silk. Fantastic demand developed for nylon as a replacement for women's silk stockings because military demand had exhausted the supply during World War II. Not until the next decade was nylon introduced as an automotive material. The first use was as a dome light lens. Its snap-fit properties were quickly recognized.

The gradual addition of rubber to seal the body provided the automobile with almost living-room comfort. Without the development of these elastomers, it is very doubtful that the smooth, quiet car of today would be possible.

With styling now selling cars more and more, bright trim was being specified. Ford began to use stainless steel for radiator shells, cowl moldings, hubcaps, door handles, and lamps on the Model A. But lower-cost chrome plating was a successful competitor.

The 1930s became the start of the chrome age; however, its birth actually occurred in 1927 when General Motors replaced nickel trim with chrome, and the new General Motors Art and Color Studio under Harley Earl produced as its first design the LaSalle. This craze for chrome did not climax until 30 years later.

Vivid, intense, authentic in its application of curite hues

La Salle Roadster
BODY BY FISHER

Harley Earl and the General Motors Art and Color Studio combined chromium plating and Duco lacquer to create this spectacular LaSalle.

The Modern Era: 1939 to 1949

In 1940, the first section of the Pennsylvania Turnpike was completed between Irwin and Middlesex. The cars of the late 1930s and 1940s can truly be described as modern transportation. Even today, the autos of this period could be driven comfortably and dependably coast to coast at conventional speeds on today's interstate system.

During this period, rayon replaced cotton cord in tires. This was a significant and profound change to the usability of the automobile. It allowed safe, high-speed motoring without a great fear of tire failure.

Plastics as materials were in style. The 1941 Packard boasted of a new multi-tone "Beautility" instrument panel in a choice of handsome, colored plastics.

World War II brought both a cessation of auto production and events that even now influence the auto industry. An SAE War Engineering Board was established. With many sources of alloying elements now cut off, SAE Iron and Steel Division, in cooperation with the American Iron and

Steel Institute, developed the National Emergency Steels, which conserved the alloying elements. In other materials areas, SAE standards were modified to conserve copper, magnesium, and aluminum.

At the same time, on the other side of the Atlantic, shell molding was being developed by Johannes Croning in Hamburg, Germany. It was not until 1947 that this significant new technique was publicized in the United States.

Copper was also in extremely short supply in Germany. This forced the development of alloys of zinc and aluminum as substitutes for bronze. This white bronze eventually led to the ZA alloys used in sleeve bearings.

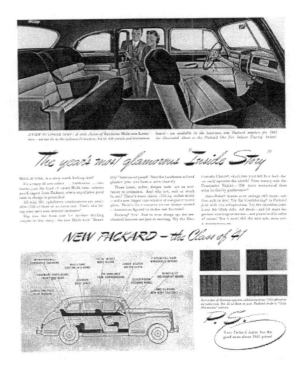

The 1941 Packard featured a new plastic multi-tone "Beautility" instrument panel.

The Jukebox Era: 1950 to 1960

In late 1951, the American automobile industry built its 100-millionth car, only a little more than 50 years after it came into existence. With America's manufacturing muscle supreme in the world and prosperity arriving, the car-buying public settled in for a return to a glitzy version of the early 1930s horsepower race—more cast iron for large displacement

The Automobile: A Century of Progress

V8s, more steel for frames and bodies, more glass for windshields. Styling was king. The same materials were being used, but in larger, more sophisticated packages.

The public seemed to want more of the same. Larger and larger cars with tacked-on chrome overshadowed more important advancements in driver comfort brought about by automatic transmissions, power-assisted brakes and steering, and air conditioning. By 1958, the height of the chrome craze, 8% of the autos sold in the United States were small, imported models.

From the materials standpoint, two developments had far-reaching implications. The first development was and is well connected to the auto industry. In 1953, General Motors introduced the Corvette with a fiber-reinforced thermoset plastic body. Almost at the same time, Kaiser introduced a reinforced-plastic-bodied sports car. Ideal for low-production runs and light in weight, the new material became popular. Even Packard used a fiberglass hood scoop and bumper shields on its low-volume Caribbean.

By 1955, the average automobile contained approximately 5 kg (11 lb) of plastic. Much of it was used in decoration: trunk lock covers, hood ornaments, horn buttons, radio dials, and wheel cover ornaments. Functional parts such as nylon gears had been used since 1940 and phenolic water-pump impellers since the 1920s. The idea of using plastic as a body material was turning practical. Henry Ford's "soybean plastic" trunk lid was becoming a reality.

Plastics for body components offered the industry new freedom of choice. The new material was light, strong, did not rust, and could be molded into complex shapes. However, problems existed: the material had excessive shrinkage, and slow process cycles required hand finishing to reach a paintable stage. Nonetheless, plastic allowed low-volume runs and styling not afforded by other materials. The competition had begun.

Unfortunately, similar to die castings decades earlier, plastics were misapplied. Poor plastics design and immature usage caused a poor image to develop. The glamour that plastics had in the 1930s was tarnished, but the advantages of plastics were too great to ignore and thus development continued.

The other materials development centered on materials science not related to the automobile industry, but it would ultimately have an almost fantastic impact on the concept of an automobile. This is really no different than many other materials developments conceived for a different purpose but ultimately adapted to the automobile.

At Bell Laboratories on December 23, 1947, using a crystal of germanium and some gold foil, two researchers were able to magnify a signal 50-fold through the crystal. Thus, the transresistor, or transistor as it is called today, was born. This piece of germanium was a semiconductor, one of a family of materials that, because of its crystal lattice, acts as a conductor of electricity in one circumstance and as an insulator in another.

The vacuum tube, invented in 1906 by Lee DeForest, was made obsolete. The switching, amplifying, and rectifying duties of the large, power-hungry, and short-lived vacuum tube were replaced by the transistor.

In 1954, two Bell scientists, Carl Frosch and Lincoln Derick, accidentally discovered that silicon dioxide, thinly coated on electrically conductive silicone, acted as an insulator. This fundamental materials discovery was critical in the miniaturization of electronics. By bathing the chip with hydrofluoric acid, the silicon dioxide could be etched away, allowing for the integration of circuitry.

During the vacation period of Texas Instruments in July 1958, Jack St. Clair Kilby managed to integrate transistors, diodes, resistors, and capacitors into one device—all from a piece of germanium. With most employees on vacation, he could not find a piece of silicon. The microchip was born.

How important were these simple materials inventions? Imagine an automobile of today with enough vacuum tubes and wires to handle the complex management of the powertrain, much less the creature comforts. By the end of the 1950s, the industry had moved from an essentially pre-war car to one that featured electronic ignition and transistorized radios.

Muscle Cars and Back to Sanity: 1960 to 1970

By 1960, the American auto industry was almost convinced that the public's automobile desires had changed. The excesses of the late 1950s cars, a recession, and inroads of the more efficient and compact imports had captured the public's attention.

In 1960, the car-buying public was flowing into dealers' showrooms to view the new Falcons, Larks, Ramblers, Corvairs, and Valiants. These cars did not excite the enthusiasts, but they did establish the groundwork for the emergence of a group of real American classics.

In 1963, John DeLorean introduced a big-block V8 into the compact Tempest platform. With its heat-treated, ground, and shotpeened driveshaft, it was a novelty; however, by the end of the year, the true muscle car was born.

Plastics continued on their winning ways. From 1960 to 1970, plastics use for a typical car grew from 11 to 45 kg (24 to 99 lb). Most of this mass increase went into low-stressed decorative components of the body and interior. Plastics had reached a point at which the public had accepted them in their own right, not as a cheap substitute. A plastic fender liner did not rust; an interior dash had a modern appearance.

In the 1960s, materials followed the trends set earlier, but events were taking place that would have a dramatic effect on materials development. The public and the government were realizing that something had to be done about air pollution in the nation's urban sprawl.

The requirements laid down first in the mid-1960s by California triggered funds to be allocated toward catalyst research. Although emissions requirements were met in the 1960s with minor add-ons and engine tuning, the stage was set for a revolution in both design and materials.

The Tough Years: 1970 to 1985

Today's automobile enthusiast has a poor regard and disinterest in the automobile of the 1970s and early 1980s. By and large, the cars were poor in performance and fuel economy. They seemed to be (and were) bolted-together with a lack of overall design cohesion.

Automobiles from this era became ugly as the bodies and frames rusted prematurely, and a gasoline shortage developed with an embargo on oil from the Arab nations. In January 1974, the speed limit went to 55 mph (88.5 km/h). Then, with enactment of the 1975 Clean Air Act, the auto industry faced a literal deadline.

The first automobile to meet the new standards was the 1975 Chrysler Avenger, which employed a rhodium-promoted platinum oxidation catalyst manufactured by Johnson Matthey. A year later, Chrysler borrowed technology from another industry and introduced computer control to engines to help walk the narrow line between emissions and performance.

Catalysts had been long employed in industry; therefore, a knowledge base was present in existing companies that produced catalysts. However, new materials problems had to be solved for the automobile industry. The exhaust system of an automobile is not a steady-state operation. The changes in temperature not only caused performance problems but also physical stress to the delicate catalytic surface.

The early automotive catalytic converters used either pellets or a honeycomb structure of stainless steel and ceramics. Early work showed that platinum-group metals were required. Lower-cost catalytics did not survive at the higher temperatures or were destroyed by sulfur in the combustion mixture. This problem was compounded by the need for a higher light-off temperature with less precious metals.

It is easy to see why platinum was chosen for the first converters. It is one of the heaviest metals known, can be readily worked, does not combine readily with oxygen or sulfur, and has a melting temperature of 1772°C (3222°F). Unfortunately, it is also very expensive, at more than $13/g today.

During the 1970s, the only available source for the metal in quantity was South Africa. No doubt there were some apprehensive people in the materials and purchasing groups.

Most of the other converters that appeared on the market used platinum and palladium. This was no small problem for the industry. However, with an air pump to deliver enough oxygen to the catalyst, combustion was

LIGHTWEIGHT MATERIALS KEY TO CHRYSLER'S CAFE

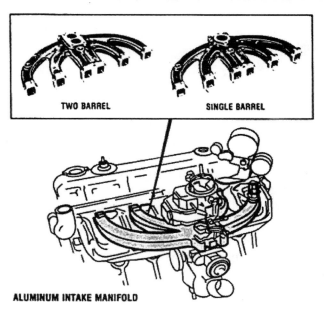

TWO BARREL   SINGLE BARREL

ALUMINUM INTAKE MANIFOLD

By the end of the 1970s, engineers were back to cutting weight.

sufficient to help regain engine performance and economy. Approximately 31 million g/yr (1.1 million oz/yr) were being used for U.S. autos from a total platinum group production by South Africa and the Soviet Union of 124 million grams (4.4 million oz).

To allow the body to have equivalent life to that of the powertrain, U.S. Steel started treating steel for corrosion protection, thus freeing the paint coatings from bearing the burden of protecting body surfaces. In the mid-1970s, Porsche introduced zinc coatings on body panels that eliminated the spangles that interfered with paint finishing.

By the end of the 1970s, engineering journals were full of new designs for weight savings, driven by the need to meet CAFE (Corporate Average Fuel Economy) standards. Not since the early part of the century had so much attention been paid to lightweight materials such as aluminum, plastics, and thin-wall zinc for weight savings. Steel frames and bodies were being put on a diet for the sake of fuel economy and performance. The stage was set. The events of these difficult years provided the incentive for producing today's long-lived, fuel-efficient, and clean-burning cars.

Computer Efficient: 1981 to 1996

The past 15 years have seen the integration of not only components, but also of materials to the whole car. No longer do automotive materials refer only to iron, steel, aluminum, and rubber. The cars of the past 15 years carry fiberglass circuit boards, gold plating, and zirconia oxygen sensors. New materials are filling roles that were only dreams 25 years ago.

With respect to the traditional automotive materials, Dan Holt in a recent *Automotive Engineering* editorial, aptly described the situation: "The battle continues. A typical family vehicle in 1995 has been estimated to contain 67.5% steel and iron, 7.7% plastics, and 5.8% aluminum." Steel and iron are being pushed by the lighter materials. Fiberglass suspension springs are now a reality. What steel loses to aluminum and composites, it gains in high-strength door components.

Now consumers are demanding a maintenance-free exhaust system—and that means stainless. While steel weight may trend downward, its value is rising. Stainless steel and ceramics compete for the converter market. Ceramic honeycomb offers better cost and lower thermal expansion. Stainless offers less thermal mass for faster heating. As electrically heated converters are developed, efficient combinations of these materials will be utilized. There

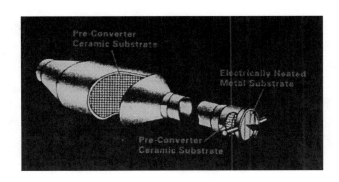

An electrically heated preconverter: materials and design continue to work in partnership during the 1990s.

is a never-ending effort to lower the cost and lessen the potential scarcity of the catalysts. Palladium, at a third the cost of platinum, is increasingly being used.

Even iron has its high-tech material: compacted graphite. Although its potential has not been fully tapped, it offers a vermicular graphite shape rather than flake, thus doubling the fatigue life and making possible lightweight, high-strength iron castings for use as cylinder heads and manifolds.

As designers improve their understanding of powder metals properties, they are increasingly applying them. Because of their narrow mass and dimensional tolerances, powder metal forgings are taking over connecting-rod applications.

In 1936 Volkswagen adapted magnesium for a reason. Today the world's manufacturers are increasing magnesium use for the same reason: weight reduction. We have returned to the basics that were understood before World War I, when automotive designers used aluminum and high-strength steel to reduce weight and improve efficiency.

Reduction in weight helps performance and makes emissions attainment easier, which in turn improves fuel economy. Now, more than ever, materials research and development is critical for satisfying customers.

Now, as it was in the dim beginning of this industry, the role of those individuals working on materials is often backstage. The automobile designer will always be creative. Many ideas and designs have been spawned, only to wait until materials development turned dreams into reality.

What is the most important materials development in the past 100 years? My vote is for the materials development that led to the modern rubber tire. Can it be imagined how far a modern auto would travel with the tire materials used by Charles Duryea on his weekend journey of long ago? What is your vote?

About the Authors

Tom Asmus
Senior Research Executive,
Technical Affairs
Chrysler Corporation
(Chapters 1, 2, and 3)

Since 1973, Tom Asmus has been employed at the Chrysler Corporation. There he has been involved with all facets of engine research, energy management systems/alternate engines, all facets of emissions control and fuel economy improvements, and fuel effects. He has been a member of the Society of Automotive Engineers (SAE) since 1976, and has participated on its Fuels and Lubricants Activity and the Passenger Car Activity/Engine Committee. He has a B.S., M.A., and Ph.D. in physical chemistry, and is a Post-Doctoral Fellow in Combustion Kinetics.

Gary Bragg
Supervising Engineer,
Engine Systems,
Mechanical Injection Group
Caterpillar Inc.
(Chapter 14)

Gary Bragg is a graduate of Bradley University School of Engineering and a 40-year employee of Caterpillar Inc., where he has been involved with basic engine, materials, and fuel-system design. Currently he is a Supervising Engineer in the Engine Systems, Mechanical Injection Product Group.

Mr. Bragg is a lifelong automobile collector and is Vice President and Director of Wheels O' Time Museum in Peoria. He is a member of the SAE Historical Committee and the Diesel Fuel Injection Equipment Standards Committee.

Carl W. Cowan
(Chapter 4)

Carl W. Cowan has considerable experience in the automotive brake community. An SAE member for almost 50 years, Mr. Cowan has served on the SAE Brake Committee for 15 years. Having received a B.S.M.E. from the Lawrence Institute of Technology (Highland Park, Michigan) in 1940, Mr. Cowan first worked for Timken Detroit Axle (heavy-duty truck and bus air brakes) until 1949. From 1949 to 1980, he worked for Bendix (automotive brakes and friction materials). From 1980 until his retirement in 1991, he worked for Abex, concentrating on brake friction materials (non-asbestos type).

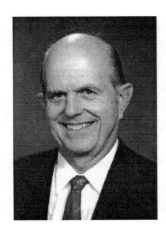

Lewis M. Gibbs
*Fellow, Transportation Fuels
Performance Unit,
Fuels and Processing Technology Group
Chevron Research and Technology
Company*
(Chapter 12)

After receiving his B.S. and M.S. degrees in mechanical engineering from the University of California, Lewis M. Gibbs has been involved with fuel research and development for more than 37 years at Chevron Research and Technology Company in Richmond, California. Currently, Mr. Gibbs is a CRTC Fellow in the Transportation Fuels Performance Unit of the Fuels and Processing Technology Group. He represents Chevron on various committees of the American Petroleum Society (API), American Society for Testing and Materials (ASTM), Coordinating Research Council (CRC), National Conference on Weights and Measures Petroleum Subcommittee, and Society of Automotive Engineers (SAE).

Mr. Gibbs is an SAE Fellow and chairman of the SAE Fuels and Lubricants Technical Committee 7—Fuels and has served as a member of the SAE Board of Directors and of the SAE Historical Committee. In ASTM, Mr. Gibbs serves as chairman of the specifications section of Subcommittee D02.A on Gasoline and Oxygenated Fuels. He is the chairman of the CRC Performance Committee.

Mr. Gibbs is the author of many articles and technical papers on fuels, of which two were SAE historical papers. He is also the author of a chapter in a new encyclopedia on energy and the environment.

Ralph H. Johnston
(Chapter 11)

Ralph H. Johnston retired in 1996 as Manager of Future Systems, Indianapolis Technical Center, Delphi Energy and Engine Management Systems (formerly Delco Remy), a division of General Motors. He was responsible for R&D in automotive-related electrical and control systems. He joined Delco Remy in 1960 and spent nearly 36 years in advanced automotive electrical and electronic systems development, including work on hybrid and electric vehicles.

Prior to joining Delco Remy, Mr. Johnson worked in the development of gyroscopic instruments for the aircraft and aerospace industry. He graduated in 1949 from the RCA (Radio Corporation of America) Institute's Electrical/Electronics Engineering program.

Karl E. Ludvigsen
(Chapters 7 and 8)

Karl E. Ludvigsen has a record of accomplishment at senior levels in the world motor industry and has also received wide recognition for his work as an editor, journalist, and author.

In 1983, Mr. Ludvigsen founded Ludvigsen Associates Limited. Since then, the company has become the leading independent European management consultancy dedicated specifically to the motor industry. Mr. Ludvigsen is also managing director of Euromotor Reports Limited, founded in 1989. Euromotor Reports is well known and respected in the world motor industry for its in-depth studies of industry topics and issues.

Before forming Ludvigsen Associates, Mr. Ludvigsen was a Vice President of Ford of Europe and a member of the Supervisory Board of Fordwerke AG of Cologne, Germany. His previous positions included a Corporate Vice Presidency of Fiat Motors of North America, public affairs responsibilities at General Motors, and the editorship of *Car and Driver* magazine.

Mr. Ludvigsen is the author, co-author, or editor of 25 books about cars and the motor industry and the winner of numerous awards for his work as an author and historian. As a designer at General Motors Styling Staff, Mr. Ludvigsen planned experimental front-drive prototypes. He also worked on heavy-duty truck transmissions in the engineering and experimental departments of the Fuller Manufacturing Company.

Mr. Ludvigsen has been a member of SAE since 1960 and is a member of the Vehicle Configuration Committee.

Larry M. Rinek
Senior Consultant
SRI International
(Chapter 4)

Larry M. Rinek, senior consultant at SRI International (Menlo Park, California), has worked with the transportation equipment industries for more than 20 years at SRI. Mr. Rinek earned a B.S. (*cum laude*) in industrial engineering in 1969 and in 1971 an M.B.A. in marketing, both from UCLA. An SAE member since 1990, he served as chairman of the SAE mid-California section and as chairman of the SAE Historical Committee. Mr. Rinek has authored 10 historical publications, including three SAE technical papers, plus seven articles in various journals concerning early American developments in aircraft, engines, motorcycles, and automotive technology. In regard to automotive brakes, he has long-term personal experience operating many types in passenger cars, light trucks, motorcycles, aircraft, and go-carts!

Gordon L. Rinschler
Executive Engineer,
Powertrain Engineering
Large Car Platform,
Chrysler Corporation
(Chapters 1, 2, and 3)

Gordon L. Rinschler was a charter member of Chrysler's Large Car Platform team, joining when it was formed in December, 1988, as Executive Engineer, Large Car Platform Engine Engineering. He is currently responsible for design and development of powertrains for Chrysler's Cirrus and Stratus, as well as all LH passenger cars. His experience with Chrysler Corporation has also included Chief Engineer, Powertrain Programs; Chief Engineer, Advanced Product Planning; Manager, Engine Design and Development; Manager, Advanced Engine Systems; Supervisor, Advanced Engine Design; Product Development Specialist, Engine Design; Research Engineer, Power Plant Research; and Project Development Engineer, Engine Design.

Mr. Rinschler's educational background includes a Professional Degree in Management, and an M.S.E. in Mechanical Engineering from the University of Michigan in Dearborn, an M.A.E. in the Chrysler Institute Program, and a B.S.M.E. from Bucknell University.

An active SAE member, Mr. Rinschler co-authored SAE Paper 840252, *Turbocharging the Chrysler 2.2-L Engine* (1984) and was program co-chairman of the 1989 SAE government/industry meeting.

Syed Q.A. Rizvi
Senior Research Chemist
The Lubrizol Corporation
(Chapter 13)

Mr. Rizvi has a Ph.D. in chemsitry and a master's degree in business. He has worked in research and development at The Lubrizol Corporation for more than 16 years and currently holds the position of Senior Research Chemist. He has authored or coauthored many publications and patents.

William J. Woehrle
President
Automotive Engineering Management
Services, Inc.
(Chapters 5 and 6)

William J. Woehrle spent 25 years with Uniroyal and the Uniroyal Goodrich Tire Company. Beginning in 1966 as a Development Engineer in the Advanced Tire Products Section, he progressed through several assignments including Resident Engineer at the tire company's proving ground in Laredo, Texas. He became Director of Tire Evaluation and was responsible for all test facilities and test operations in support of automotive original equipment qualifications activities.

Mr. Woehrle is a past president of the Tire and Rim Association and past chairman of the Tire Engineering Policy Committee of the Rubber Manufacturer's Association. He also served as chairman of the SAE Highway Tire Forum Committee.

In addition, Mr. Woehrle is President of Automotive Engineering Management Services, Inc. (AEMS). This firm consists of an assembly of former engineers and scientists from Uniroyal and B.F. Goodrich, together with experts from the automotive and aerospace industries. It specializes in independent automotive testing services and evaluations, and is located in Brighton, Michigan.

Anthony J. Yanik
(Chapters 9 and 10)

Anthony J. Yanik, now retired, was an employee at General Motors Corporation's Automotive Safety Engineering Center since joining the company in 1968. He specialized in the writing of features and corporate speeches on automotive safety subjects and was the corporate resource on matters pertaining to the aging driver.

A member of SAE, he originated and headed the Mature Driver Standards Committee and has presented numerous SAE papers on this subject, including SAE 851688, *What Accident Data Reveal About Elderly Drivers*; SAE 870237, *How Aging Affects the Relationship Between the Driver and the Road Environment*; and SAE 900192, *New Technology Considerations for Mature Drivers*. He also organized and chaired a special session in 1988 honoring five esteemed automobile designers of historical import, such as Gordon Buehrig, Zora Arkus-Duntov, and Robert Bourke. In 1992, he presented SAE 920845, *The Automobile: Unwanted Technology—The Later Years*. He serves on the SAE Historical Committee.

Index

B

C

D

G

I

J

K

L

Rubber *(continued)*
    synthetic, 58
    vulcanization of, 56
Rubber materials, 237, 241

S

SAE Handbook, first, 235
Safety
    interior, 116-119
    test technology, 140
    See also Transportation safety
Safety glass, 98
Safety rim wheel, 131
Saturn, 92*p*
SBR. *See* Styrene butadiene rubber
Schrader, George H., tire inflation/deflation and, 68
Sealed beam headlamp, 130-131
Seat belts
    mandated use of, 144, 147
    as safety feature, 135-136
Seating
    from 1896 to 1914, 100
    from 1918 to 1933, 104-105
    from 1934 to 1941, 107
    from 1945 to 1956, 111
    from 1957 to 1967, 113*p*, 113-114
    from 1968 to 1980, 116
    from 1981 to 1995, 119
    adjustable, 111
    padded, 131
Self-adjusting drum brakes, 135
Sheet-steel press tool, all-steel bodies and, 83
Sheffield Simplex, interior of, 102*p*
Shoulder belt, 137
Side glass, curved, 98
Side windows, 100
Side-Guard door beam, 140
Side-impact protection, 148
Sidewall styling, 67-68
"Silvertown" tire, 53

T

U

DATE DUE

May 30/02				
Oct. 24/12 Jan 30/17				
May 30/17				